How to Go Digital

The Digital Future of Management
Paul Michelman, series editor

How to Go Digital: Practical Wisdom to Help Drive Your Organization's Digital Transformation

What the Digital Future Holds: 20 Groundbreaking Essays on How Technology Is Reshaping the Practice of Management

How to Go Digital

Practical Wisdom to Help Drive Your Organization's Digital Transformation

MIT Sloan Management Review

The MIT Press
Cambridge, Massachusetts
London, England

This book was set in Neue Haas Grotesk and Stone Serif by the MIT Press. Printed and bound in the United States of America.

Library of Congress Cataloging-in-Publication Data

Names: Sloan Management Review Association, compiler.
Title: How to go digital : practical wisdom to help drive your organization's digital transformation / MIT Sloan Management Review.
Description: Cambridge, MA : MIT Press, [2018] | Series: The digital future of management | Includes index.
Identifiers: LCCN 2017027097 | ISBN 9780262534987 (pbk. : alk. paper)
Subjects: LCSH: Business--Data processing.
Classification: LCC HF5548.2 .G594 2018 | DDC 658/.05--dc23 LC record available at https://lccn.loc.gov/2017027097b

10 9 8 7 6 5 4 3 2 1

Contents

Series Foreword ix

Introduction: Digital Transformation Might Be Different Than You Think xi
Gerald C. Kane

I Laying the Foundation 1

1
How to Develop a Great Digital Strategy 3
Jeanne W. Ross, Ina M. Sebastian, and Cynthia M. Beath

2
Five Myths about Digital Transformation 13
Stephen J. Andriole

3
Winning the Digital Talent War 19
Gerald C. Kane, Doug Palmer, Anh Nguyen Phillips, and David Kiron

4
Reframing Growth Strategy in a Digital Economy 29
Didier Bonnet and Pete Maulik

II Building Data-Based Value 37

5
How to Monetize Your Data 39
Barbara H. Wixom and Jeanne W. Ross

6
What's Your Data Worth? 49
James E. Short and Steve Todd

7
Is Your Company Ready for HR Analytics? 59
Bart Baesens, Sophie De Winne, and Luc Sels

8
Why Your Company Needs Data Translators 65
Chris Brady, Mike Forde, and Simon Chadwick

III Operational Makeovers 73

9
Sales Gets a Machine-Learning Makeover 75
H. James Wilson, Narendra Mulani, Allan Alter

10
A New Approach to Automating Services 81
Mary C. Lacity and Leslie P. Willcocks

11
Organizing for New Technologies 105
Rahul Kapoor and Thomas Klueter

12
Mastering the Digital Innovation Challenge 111
Fredrik Svahn, Lars Mathiassen, Rikard Lindgren, and Gerald C. Kane

IV New Approaches to Social Media 121

13
Finding the Right Role for Social Media in Innovation 123
Deborah L. Roberts and Frank T. Piller

14
Improving Analytics Capabilities through Crowdsourcing 141
Joseph Byrum and Alpheus Bingham

15
Beyond Viral: Generating Sustainable Value from Social Media 155
Manuel Cebrian, Iyad Rahwan, and Alex “Sandy” Pentland

16
When Employees Don’t “Like” Their Employers on Social Media 159
Marie-Cécile Cervellon and Pamela Lirio

V Diving into the Void 179

17
Leading in an Unpredictable World 181
Pierre Nanterme, interviewed by Paul Michelman

Contributors 193

Index 199

Series Foreword

Books in the Digital Future of Management series draw from the print and web pages of *MIT Sloan Management Review* to deliver expert insights and sharply tuned advice on navigating the unprecedented challenges of the digital world. These books are essential reading for executives from the world's leading source of ideas on how technology is transforming the practice of management.

Paul Michelman
Editor in chief
MIT Sloan Management Review

Introduction: Digital Transformation Might Be Different Than You Think

Gerald C. Kane

People throw around the term "digital transformation" with ease these days, but there isn't much agreement on what it actually means. Originally, the value in the phrase was that it conveyed the need to engage in a fundamental shift in the way we think, work, and manage our organization in response to digital trends in the competitive environment. While the need for fundamental change remains, the overuse and, often, misuse of the term has weakened its potency.

What Is Digital Transformation?

Perhaps the most common definition—that digital transformation is about the implementation and use of cutting-edge technologies—is also the most misguided. It's not hard to find a company that has deployed a new digital tool or platform just to have it remain unused by employees or unable to deliver the intended transformative impact on the business.

Another interpretation of digital transformation is that it involves organizations using technology to do business in new and different ways. This definition is certainly better, but it remains incomplete. For example, many companies are adopting new talent models in response to digital trends. Employees follow two- to three-year "tours of duty," engaging in one project or role for a certain period of time, at which point they transition to a new role in order to develop different skill sets continually. These efforts are clearly and intentionally designed to allow a company to cultivate diverse talent in a rapidly changing digital world, but it doesn't involve implementing or using new technology at all.

I believe the most productive view of digital transformation is that it is about adopting business processes and practices that position organizations to compete effectively in an increasingly digital world. This definition of digital transformation has three important implications for managers.

First, it means that, fundamentally, digital transformation is about how your business responds to digital trends that are occurring whether you initiated them or not—and whether you like them or not. Much of the need for digital transformation is outside your control. It is about adapting to how your customers, partners, employees, and competitors use digital technologies—and how that use changes their behavior and expectations.

Second, it means that how an organization implements new technologies is only a small part of digital transformation. In cases where transformation does directly involve new technologies, the tool is only part of the story. Other issues, such as strategy, talent management, organizational structure, and leadership, are equally or more important than the technology itself.

Third, it doesn't require special knowledge or special skills to be a leader in the digital age. As the world is transformed by digital tools, all managers are digital managers by default. But there is a mighty gulf between those who are effective digital leaders and those who are not—and that gulf is measured by one's willingness to transform and adapt.

What Does It Take to Be an Effective Digital Leader?

In the 2016 *MIT Sloan Management Review* report on digital business, "Aligning the Organization for Its Digital Future," we asked business executives to name the most important skills a leader needs to succeed in a digital environment. The responses were far more managerial than technical in nature: strategic vision, forward-looking perspective, change-oriented mindset—categories that encompassed 80 percent of the responses.

Only 20 percent of respondents indicated that technical skills were the most important trait of a digital leader. Moreover, the technical skills that they indicated *were* important involved general technological capacity rather than hard-core technical chops. Leaders need a solid understanding of digital trends and capabilities to develop a robust strategic vision for a digital environment, to maintain a forward-looking perspective, and to be able to identify the changes necessary to adapt to a continually changing competitive environment. And those are precisely the competencies this book aims to help leaders build.

Our ongoing study of digital business at *MIT SMR* shows that having a clear and coherent digital strategy is the most important factor associated with digitally mature organizations. Several articles in this collection will help put you on the right strategic path. "How to Develop a Great Digital Strategy" will help you begin thinking about how your organization should compete, while "Reframing Growth Strategy in a Digital Economy" will guide your understanding of what to expect when you start implementing that strategy. "How to Monetize Your Data" looks at key strategic questions with respect to analytics, and "What's Your Data Worth?" will help you analyze the value that may be locked up in the data your company possesses. "Beyond Viral: Generating Sustainable Value from Social Media" invites similar strategic questions in relation to social media technology.

Our research further suggests that talent acquisition and retention are critically important but often overlooked facets of digital transformation, and this volume addresses these issues as well. "Is Your Company Ready for HR Analytics?" asks how data can be used to help support effective workforce decisions, and "Why Your Company Needs Data Translators" suggests what talents you may need to make good use of that data. "Improving Analytics Capabilities through Crowdsourcing" challenges managers to use digital platforms to bring outside expertise into an organization. "When Employees Don't 'Like' Their Employers on Social Media" discusses how to use similar digital platforms to enable your employees to go outside your organization and become brand ambassadors.

Last, you will find articles that focus on the need to organize differently as a result of digital trends. "Winning the Digital Talent War" asks whether and how the company–employer relationship will evolve as a result of digital technologies. "Organizing for New Technologies" discusses ways that companies can organize to make them more amenable and adaptable to digital innovation, while "Mastering the Digital Innovation Challenge" suggests that one set of changes often cascades to require other sets of changes in order to fully embrace digital innovation capabilities. "Finding the Right Role for Social Media in Innovation" explores different ways to use social platforms to enable innovation.

Noteworthy is that none of these articles is strictly about implementing new digital technologies. Instead, they are about how to adapt your organization to respond to and take advantage of opportunities in the competitive environment made possible by digital technologies. Of course, these opportunities and threats will continue to change as digital technologies do. Digital leaders will need to navigate their organization amid ongoing and unexpected changes. Chapter 17, "Leading in an Unpredictable World," helps you think about this reality.

Where Do You Start with Digital Transformation?

Appropriate knowledge and skills are the necessary building blocks for digital transformation, but they won't get the job done alone. You must also have the will to lead.

Organizations must make a conscious effort to transform. Executives indicate that a major barrier to effective digital transformation is simply that an organization has too many competing agendas. Only by highlighting digital transformation as an effort of paramount importance will an organization prioritize it appropriately. Some organizations have even begun identifying a chief transformation officer, who is responsible for spearheading major digital change efforts.

The time for transformation is now. The gap between what digitization may promise (or threaten) and how companies continue to operate is growing wider. If your organization waits until evidence from the marketplace proves your legacy business model is failing, it will be too late. Digital transformation requires rethinking one's entire business, step-by-step, from the ground up, over time.

Successful digital transformation is not a one-and-done effort. It requires a flexible mindset and an organizational structure that allows a company to respond to digital trends continually. As such, digital transformation is an ongoing process. A common mantra is to "fail fast" as your organization experiments with new digital business processes. This is solid advice, but the actual problem may be the opposite: Many companies don't spend enough time thinking about what to do if they actually succeed, nor about how to use such successes to drive fundamental change across the enterprise.

Last, marry your pain points to your vision of a new state. Engage in a careful analysis of your organization's current operations as compared with a digital ideal. This will help you identify where to start. It is not sufficient for leaders simply to say the organization needs to be more agile. Rather, you must carefully articulate how the organization needs to become more flexible, resilient, and adaptive. What factors are preventing the emergence of those characteristics?

The good news is that getting started is the hardest part. Many companies report that once they make the first few steps toward becoming a more digital enterprise, their efforts begin to gain momentum. A few small successes can help the organization begin to see and share your vision of what is possible and become more willing to actively help bring it about. This book is full of ideas to help your organization take the first—or next—step toward digital transformation. What are you waiting for?

I

Laying the Foundation

1

How to Develop a Great Digital Strategy

Jeanne W. Ross, Ina M. Sebastian,
and Cynthia M. Beath

As leading technology companies embrace biometrics, artificial intelligence (AI), drones, and other exciting digital technologies, senior business executives at many other companies feel pressured to do the same. But if they are to maximize the value from investment in new technologies, business leaders first must make sure that their companies have a great digital strategy.

We studied digital strategies as part of a research project on designing digital organizations that the MIT Center for Information Systems Research conducted in partnership with The Boston Consulting Group; in that project, we interviewed more than 70 senior executives at 27 companies. Our findings underscored the importance of developing a winning business strategy that takes advantage of digital technologies. A great digital strategy provides direction, enabling executives to lead digital initiatives, gauge their progress, and then redirect those efforts as needed. The first step in setting this direction is to decide what kind of digital strategy to pursue: a customer engagement strategy or a digitized solutions strategy.

A customer engagement strategy targets superior, personalized experiences that engender customer loyalty. A digitized

A great digital strategy provides direction, enabling executives to lead digital initiatives, gauge their progress, and then redirect those efforts as needed.

solutions strategy targets information-enriched products and services that deliver new value for customers. The best strategy for a company will depend on its existing capabilities and the way it wants to compete. The most important requirement for a great digital strategy, however, is to choose one kind of strategy or the other, not both. A digital strategy aimed at operational excellence may appear to be a third choice, but increasingly, operational excellence is the minimum requirement for doing business digitally, not the basis for a sustainable competitive advantage.

Customer Engagement Strategies

The focus of a customer engagement strategy is the development of customer loyalty and trust—and, in the best cases, passion. Companies choosing this approach offer seamless, omnichannel customer experiences, rapid responses to new customer demands, and personalized relationships built upon deep customer insights. Recognizing the always-rising bar of customer expectations, companies with a great customer engagement strategy are constantly identifying new opportunities to connect with their customers.

Kaiser Permanente, an integrated provider of health care and not-for-profit health plans based in Oakland, California, is following a customer engagement strategy. Driven by what it calls its "consumer digital strategy," Kaiser Permanente approaches health care as a collaboration between care providers and members. It uses digital technologies to provide seamless, low-cost access to provider teams and to facilitate the delivery of both curative and preventive patient care.

Kaiser Permanente capitalizes on digital technologies by:

Offering digital channels that bolster patient interaction with care delivery teams. Kaiser Permanente's channels provide access to personal health records, secure messaging between patients and providers, and remote care.

Applying data analytics to identify the need for – and the most effective approach to – personalized medical outreach. Kaiser Permanente uses analytics to track and improve patient compliance with medication and treatment regimens, and to identify the most effective forms of outreach for generating healthy behaviors.

Leveraging social media to develop communities of patients with similar interests and needs, and to create "care circles," where patients and their families can engage with care providers. Kaiser Permanente is using a carefully crafted permission system to allow approved family members and other caregivers to help support patients, communicate with their physicians, and monitor their treatment.

As a result of these efforts, 70% of Kaiser Permanente's members are actively engaged in managing their health online—a behavior that the organization's research indicates is positively correlated with better member health, adherence to treatment regimens, satisfaction, and retention.

Digitized Solutions Strategies

A digitized solutions strategy transforms what a company is selling. It seeks to integrate diversified products and services into solutions, to enhance products and services with information and expertise that help solve customer problems, and to add value throughout the life cycle of products and services. Over time, digitized solutions can transform a company's business model by shifting the basis of its revenue stream from transactional sales to sophisticated, value-laden offerings that produce recurring revenue.

Schindler Group, a global provider of elevators, escalators, and related services based in Ebikon, Switzerland, is pursuing a digitized solutions strategy. The company leverages the Internet of things—collecting real-time data from its installed base of elevators and escalators and using that data to improve the quality of its products and services. Initially, Schindler focused on operational excellence in its digital strategy—using analytics to reduce the costs it incurred in servicing its products. But all four global competitors in the industry are improving their operations in this way. So, Schindler shifted its focus to digitized solutions and began using data from the Internet of things to help prevent equipment failure, optimize elevator routes, and identify potentially valuable innovations.

Schindler has developed a new urban mobility solutions strategy that takes advantage of its ability to manage the efficient movement of thousands of people within a building. For example, the company has developed a digitized solution that allows visitors to bypass lobby security stations by swiping their mobile phones at the point of entry.

Choose Only One Strategy

Although customer engagement and digitized solutions strategies are very different digital strategies, their paths will invariably converge over time. For example, Kaiser Permanente's pursuit of customer engagement has led to the development of remote monitoring services, a digitized solution. Similarly, Schindler is deploying a mobile app that communicates the real-time status of elevators and escalators to facility managers, enhancing customer engagement. But this convergence doesn't obviate the need to choose between the two types of digital strategies.

Companies need a clear digital strategy to develop an integrated portfolio of customer offerings. Their employees need a clear strategy to guide their innovation initiatives and resolve debates over priorities. Thus, the strategy predetermines the winner in stalemates between product and customer factions within a company, and it discourages functional silos from pursuing independent goals.

For example, financial services company USAA, based in San Antonio, Texas, has generated outstanding net promoter scores by restricting the development (and acquisition) of financial products to those that can be integrated into a seamless customer experience. The company's customer engagement strategy drives the innovation agenda, and its digitized solutions must conform to that strategy. At Apple Inc., in contrast, a digitized solution strategy drives innovation. The company expects customer-facing employees to deliver a great customer experience, but the product comes first, even if that means employees must convince reluctant customers that they don't need a headphone plug. At both USAA and Apple, the choice of digital strategy provides a shared understanding of business objectives among employees, and it guides experimentation and innovation.

Build an Operational Backbone

To exploit the numerous opportunities for delivering on either type of digital strategy, a company also needs an integrated

A company that lacks enterprise capabilities will not be able to deliver reliable operations and thus will not be able to compete digitally.

platform of distinctive capabilities—we call it an operational backbone—that ensures efficient, reliable transactions and customer interactions. These capabilities vary by company, but they typically include things like access to a single authoritative source of information for key data about finances, customers and products; reliable end-to-end global supply chain processes; or back office shared services.

Companies have been trying—and struggling—to build such operational backbones since the 1990s, when they started implementing enterprise resource planning (ERP) software and customer relationship management (CRM) systems. In many cases, organizational politics, cultures, and processes reinforced existing business silos, which hindered implementation of these systems and the accompanying enterprise-wide capabilities. But a company that lacks enterprise capabilities will not be able to deliver reliable operations and thus will not be able to compete digitally.

Kaiser Permanente's operational backbone starts with its electronic health records system. Kaiser Permanente committed early on to electronic health records as the basis for clinical record keeping and collaboration. Now, its health records system facilitates increasingly meaningful patient interactions, and it enables new digital initiatives that require accurate, accessible patient data.

Schindler's operational backbone comprises the global business technology and process standards that it implemented with its ERP system starting in 2005. Now, as the company supplements its operational backbone with sensor data, it facilitates the company's ongoing mobility solutions experiments.

In assessing their company's ability to execute one of the two digital strategies, business executives must be mindful of the gaps in their capabilities—and then, as quickly as possible, wire them into the organization's operational backbone. To succeed in the digital economy, companies must offer a unique value proposition that is difficult for both established competitors and startups to replicate. Such a value proposition stems from a digital strategy that is focused on either a set of digitized, integrated offerings or a relationship that engages customers in ways that competitors can't match. Without that, you might create a flurry of innovations, but you won't deliver value-added applications of AI, biometrics, drones—or the next important digital technology.

2

Five Myths about Digital Transformation

Stephen J. Andriole

Many boards of directors and senior management teams aspire to the efficiencies, innovation, and competitiveness that digital transformation might deliver. But in my experience, the path to transformation—like most major corporate initiatives—is a risky one.

I have spent much of my career overseeing and participating in digital transformations in both government and private sector settings. Specifically, I have served as the director of the Cybernetics Technology Office of the US Defense Advanced Research Projects Agency (DARPA); as CTO and senior vice president of Safeguard Scientifics Inc.; and as CTO and senior vice president for technology strategy at Cigna Corp. And I have observed that in the vast majority of cases, organizations will make significant mistakes—unless the transformation is well-planned, exquisitely executed, and enthusiastically sponsored by upper management.

Villanova University—where I now teach and direct research about digital transformation and emerging technologies—collects data about technology adoption and digital transformation trends. I'm constantly hearing about the "amazing," "fabulous," "terrific," and "incredible" projects under way with the potential to "revolutionize" companies and "disrupt" whole industries. But when I probe survey respondents for key details about their initiatives, I often find that there is still confusion about the process. To replace this confusion with some clarity, I have distilled my observations and experiences into five myths about digital transformation—each of which has a corresponding reality. If you understand these myths, you'll be less likely to fall prey to the hype about digital transformation and be more aware of how arduous the process really is.

Myth #1: Every company should digitally transform.
Reality: Not every company, process, or business model requires digital transformation.

Digital transformation is not a software upgrade or a supply chain improvement project. It's a planned digital shock to what may be a reasonably functioning system. For example, to launch a digital transformation of business processes, it's necessary to purposefully model those processes with tools that enable creative, empirical simulations. Think, for example, of the software programs that enable business process modeling and business simulations.

So, as a first step to digitally transforming your processes, you need to honestly assess if your company can create digital models that simulate the nuances inherent in its procedures. Simply put, the question is this: Can my company model its existing processes? Many companies cannot. That's no crime. But that means, in all likelihood, that you cannot easily digitally transform all of those processes.

Remember, too, that the impact of any initiative is ultimately defined by market share, revenue, and profit. That means that some companies—even if they *can* model their nuanced processes—may still not be able to make a convincing business case for digitally transforming them. (In other words, just because it's *possible* doesn't mean it's going to be *profitable.*) What's more, you should keep in mind that your existing business rules, processes, models, and systems may be working just fine, so efforts to digitally transform them may not make sense, given the costs and time required of the effort.

Of course, over time, the efficiency of your rules, processes, models, and systems may diminish; when that happens, your company's need for digital transformation could grow. But you don't have to effect digital transformation just for transformation's sake; you should be able to make the business case, and you should be able to say, with certainty, that the transformation will successfully streamline some key processes.

Myth #2: Digital transformation leverages emerging or disruptive technologies.
Reality: Most short-term transformational impact comes from "conventional" operational and strategic technology—not from emerging or so-called "disruptive" technology.

Most transformational leverage comes from tried-and-true operational technology (for example, networking and databases) and strategic technology (enterprise resource planning or customer relationship management software). It rarely, in my experience, comes from emerging technology (such as augmented reality) or disruptive technology (such as machine learning).

Why is that? Many business processes and models are outdated. For example, consider the manner in which Uber Technologies Inc. and Airbnb Inc. have, by degrees, supplanted taxis and hotels respectively. While emerging technologies have abetted Uber and Airbnb's rises to prominence, their most significant gains have come from leveraging the mainstream networking technologies already in consumers' hands: mobile phones, apps, and websites optimized for quick transactions and location tracking. It's often easier to achieve impact with technologies already in widespread use than it is with emerging technologies.

As obvious as that point may seem, many leaders ignore it. They think they have to be positioned to pounce on the next wave of emerging technology, when that next wave is often difficult to predict and is, by definition, not yet conventional enough to produce a major impact.

Myth #3: Profitable companies are the most likely to launch successful digital transformation projects.
Reality: If things are going *well* – defined crassly as employee and shareholder wealth creation – then the chances of transforming anything meaningful are quite low.
Failing companies are much more motivated to transform themselves, simply because they need to change something—if not everything—quickly. Successful companies, especially if they're public companies, are understandably cautious about change. Think about it: How many successful companies—without market duress—have truly transformed their business models? Change is expensive, time-consuming, inexact, and painful. It also makes the leaders who suggest it easy targets for in-house politics, especially when the change initiatives move slowly or stumble.

And despite what the best-selling business authors, pundits, and huge-fee-collecting lunchtime speakers will tell you, the truth is, most human beings are resistant to digital change when it happens in the organization where they've grown comfortable. That means that transformation efforts are often constrained.

Yes, resistance to change can disappear quickly when a company begins to fail. But until that day arrives, it's difficult to tell everyone to fix what's perceived as unbroken.

Where is there the least resistance to digital transformation efforts? At companies hemorrhaging customers and cash, and at startups with investor cash to burn. That's because digital transformations work well when you have money to spend and a high capacity—and rationale—for taking risks. By contrast, established companies are "established" for a reason: They've reached consistent levels of profitable revenue generation, driven by well-understood processes that make up an ongoing business model. They are therefore typically unwilling to upend those processes as long as they continue winning in the marketplace.

Myth #4: We need to disrupt our industry before someone else does.
Reality: Disruptive transformation seldom begins with market leaders whose business models have defined their industry categories for years.

While market leaders pay lip service to their role as innovators and disruptors, they are usually unlikely champions of change—until their profits begin to fall and their shareholders scream for transformation.

Historically, industry disruptors have often been startups making bold bets on old industries. Examples include Airbnb (hospitality), Uber and Lyft (transportation), Amazon (books, retail), and Netflix (entertainment).

Does this mean there's no possibility for industry leaders to disrupt themselves? No. But let history serve as a helpful reminder: Disruption seldom comes from established companies with consistent, profitable revenue streams.

Myth #5: Executives are hungry for digital transformation.

Reality: The number of executives who really want to transform their companies is relatively small, especially in public companies. Digital transformation requires strong support from upper management. And while the *concept* of digital transformation can be sold up the management chain, simply selling the concept isn't enough. Transformations require overt, continuous support from the senior management team to succeed.

And it's this sort of support—public, persistent, enduring, and unwavering—that's more difficult to secure than one might assume. Many executives are suspicious of risky change efforts that might affect their status in the company. Many executives are also challenged by the sheer complexity of digital transformation projects, especially when they learn how long they take. Moreover, as we've already discussed, executives are reluctant to tweak existing business models that are consistently generating wealth for themselves and their shareholders.

In short, there's a wide gap between what executives say about digital transformation and what they do. It would be nice to think that executives are primarily motivated by what's best for the long-term health of the company, but their motives are often more complex.

3

Winning the Digital Talent War

Gerald C. Kane, Doug Palmer, Anh Nguyen Phillips, and David Kiron

As companies attempt to compete in an increasingly digital world, they face a wide range of challenges. Somehow, with limited resources and competing priorities, they must develop the capabilities they need so that their activities, people, culture, and structure are well-aligned with their organizational goals in a changing competitive environment. One of the most critical issues is finding the right people—something many companies appear to be struggling with—and designing paths forward that meet their needs. In a 2016 digital business study and research project that *MIT Sloan Management Review* and consulting firm Deloitte LLP conducted, we found that the ability of companies to attract and retain talent was one of the most serious—and most overlooked—digital threats companies faced. Seventy percent of the more than 3,700 executives, managers, and analysts we surveyed agreed that their organizations needed a new or different talent base to compete effectively in a digital world.

Yet the actual skills that organizations and their employees likely need may come as a surprise. Respondents indicated that technical skills were most important only 18% of the time for leaders and 27% of the time for other employees. Many saw other capacities—being change-oriented, forward thinking, and having a transformative vision—as equally important or even more important for working successfully in a digital environment. Today's employees are looking for opportunities to work for companies that will allow them to develop and demonstrate the skills and abilities that they need to succeed in the digital world.

Organizations that can provide such opportunities, be it through formal training or hands-on experience, have an advantage in both attracting and holding on to talent. According to our survey, respondents who felt that their employers did not offer opportunities to develop their digital skills were six times more likely to say they expected to leave the company within a year than those who worked for more digitally mature organizations where there were more skill development opportunities. Those who were disposed toward leaving weren't just the younger, less-experienced employees but also middle and upper managers, who are often seen as critical to an organization's future. Unless companies act quickly, they are likely to lose talent they currently have and experience difficulty attracting new talent.

Where Will Talent Come From?

In our interviews with digital executives and analysts, we saw two distinct approaches to thinking about talent in an increasingly digital business environment. Some companies expressed great interest in tapping into fluid talent markets made up of skilled contractors and consultants. Companies pursuing this approach seek to develop a more flexible staffing model that uses a digital platform for accessing freelance talent. For example, Work

Digital talent markets are different in that they can be used to coordinate the activities of both specialized workers and contract employees more fluidly, dependably, and in real time.

Market Inc., based in New York City, operates a freelance management platform that facilitates hiring specialized talent based on an organization's current needs. While businesses that act as brokers between companies and freelance workers have been around for many years, digital talent markets are different in that they can be used to coordinate the activities of both specialized employees and contract workers more fluidly, dependably, and in real time. Companies across many industries are using digital talent markets. For example, Olo, a digital ordering company based in New York City, is working on integrating its point-of-sale software with Uber infrastructure to provide on-demand delivery drivers for restaurants.

Talent markets based on digital platforms have been evolving in significant ways in recent years. For example, Topcoder Inc., a San Francisco–based company that connects a global community of software designers, developers, and programmers to customers, is well-known as a talent market platform. However, as the needs of the tech industry have changed, Topcoder's strategy has changed as well. Rather than just being a platform where software developers can shop for their next project, Topcoder has also begun to work with its members to develop their skills as project managers so that they can learn how to manage the on-demand talent its platform provides.

In contrast, other companies are focused on how to develop and manage existing employees for the long term. Many of these companies invest heavily in new approaches to onboarding and continuous training and development. They provide employees with opportunities to grow digitally, not only through technical training but also by offering carefully curated work experiences, different experiences over time, and career development support. Allied Talent LLC, a Silicon Valley–based consulting firm, advises companies on how to deepen employee skills while engaging employees in creating their career paths. Allied Talent recommends employees move to new projects within the organization every two to four years, taking on assignments that support both the corporate mission and the employee's career goals. This approach is designed to allow employees to develop new skills while having diverse career opportunities within the organization.

Blending Two Models

On the surface, the two approaches to attracting and managing talent in a digital world may appear to clash with each other. The first strategy encourages easy access to on-demand talent through digital platforms that can be expanded or contracted as work flows and skill-set needs change; the second calls for developing and investing in employees. Although most companies tend to emphasize one approach over the other, we suggest that a strategy that combines elements of both could provide a potentially effective way to compete in a digital world.

Such an approach would mean that digital organizations would rely on two types of talent: flexible "on-demand talent," who can be called upon when needed, and "core" employees. A division along these lines is by no means new—companies have relied on core staff and contractors to fill their talent needs for years. But digital platforms enable companies to source and organize talent more quickly and easily than ever, providing managers with new flexibility to configure staff and contractors in ways that can work best for a given project. This may require rethinking the future for both on-demand talent and core employees.

Cultivating Talent Markets

Companies may need to consider how to manage specialized, fluid talent differently than traditional employees. It may require cultivating on-demand talent markets so that specialized talent is available as needed and on demand. Talent markets can be maintained via digital platforms that monitor, evaluate, and support the talent pool of on-demand contractors.

Manage on-demand talent markets as a community. In order for companies to ensure access to the types of skill sets they need, they should recognize on-demand talent markets as strategic resources and invest in the long-term health of the talent pool itself. Individuals may come and go, but the on-demand talent market should be nurtured and maintained with an eye toward the future.

Balance full-time and part-time talent. While talent markets have typically been used to manage part-time freelancers, some companies have also begun experimenting with them as platforms for assigning full-time employees to projects as needed. For example, Work Market has set up dedicated talent pools for companies made up of both full-time employees and part-time freelancers. Full-time employees provide a stable base of employees, while part-time contractors provide the flexibility to deal with variations in demand.

For some on-demand contractors, the opportunity to become full-time employees may be a powerful motivator to continue building their skill sets. However, there are many people who are not in the market for full-time employment but still have valuable skills (for example, student workers, parents of young

Many organizations treat contractors as second-class citizens, but companies that want to attract great talent can't afford to do that. On-demand talent with valuable skills can choose to work for any project or company.

children, and people nearing retirement age). The crowdsourcing site InnoCentive Inc., based in Waltham, Massachusetts, has found that retired workers with specialized expertise are among its most valuable and regular contributors.

Create an environment where the best people want to work. Many organizations treat contractors as second-class citizens, but companies that want to attract great talent can't afford to do that. On-demand talent with valuable skills can choose to work for any project or company. To ensure that they're able to get the best, organizations should cultivate an environment and incentive structure where on-demand contractors are valued as integral contributors to the company's strategic objectives. Providing desirable work experiences and environments, opportunities to work on interesting projects, and exposure to different teams can help drive engagement.

Rethinking Core Employees

Companies that increasingly rely on these talent markets may also need to rethink the nature and roles of their full-time employees. Core employees are not just full-time employees. They are the people you plan to invest in to build and guide the long-term strategic direction of the organization. Therefore, you should think about them differently.

Train employees to delegate to on-demand talent. Although core employees will likely be working with other core employees, increasingly they may be delegating work to on-demand talent, which will require specific managerial skills. Effective delegation requires knowing how to source critical skills, how to assemble teams and get them up and running quickly, and how to use digital decision support tools effectively to meet the goals. These skills can provide the organizational agility and the collaborative environment that characterized digitally maturing companies in our survey.

Equip core employees to influence strategic decisions. Core employees, even those who are relatively junior, should have a certain level of strategic autonomy to accomplish or contribute to designated goals. Strategic thinking is one skill that respondents to our survey indicated was essential for both leaders and employees working in a digital environment, and distributed leadership was a key cultural element of digitally maturing companies. Obviously, offering greater independence would require more communication with top leadership and increased awareness of the strategic direction of the company.

Create an environment people will want to be a part of for a long time. It is no accident that a key differentiator of digitally maturing companies is the way they intentionally work to develop, maintain, and strengthen employee engagement. Keeping core employees engaged for the long term involves providing more than a paycheck. For employees to want to stay and contribute, many say they need to feel that the organization is willing to invest in them and will continue to offer opportunities for growth. The 3M Co., for example, invests in new hires to build loyalty. According to 3M CEO Inge Thulin, the company plans to put all of its employees in an expanding employee development program by 2025.

Provide diverse opportunities to gain digital experience continually. Core employees likely require new opportunities to grow their skill sets over time. Companies can create new development programs that—unlike traditional leadership development programs that selected employees take part in at certain points of their tenure—encourage core employees to continuously update their skills to stay abreast of the ever-evolving digital world.

Organizations seeking to compete using a combination of core employees and on-demand talent markets need to address some important questions. For example, how big does the core organization need to be, and what skills should the core employees have? Should companies work with existing on-demand talent markets, or should they cultivate their own to ensure that they have the right skills when they need them? How does a company build a robust on-demand talent market while keeping these skills from competitors? And to what extent is it possible to cooperate with other companies to share talent markets?

With these and other questions in mind, companies are beginning to experiment with new models for managing talent. Such experimentation could be essential for getting the most out of talent in the digital era. Talent management designed for traditional work environments may not enable organizations to compete in the digital world, and it may discourage the very people you most want to attract and retain.

The research for this article was conducted by the authors as part of a sponsored *MIT Sloan Management Review* research initiative. It was selected and edited for publication by an independent editorial team. Find the full report at http://sloanreview.mit.edu/digital2016.

4

Reframing Growth Strategy in a Digital Economy

Didier Bonnet and Pete Maulik

Digital technology is radically changing the behavior of individuals, corporations, and entire societies, and disruption seems to be the new normal. CEOs are faced with the dual challenge of protecting their backyards from upstarts and incumbents while simultaneously devising strategies that will guide their growth for the next five years.

For most, this is a daunting task. If you are a senior executive faced with these challenges, how do you ensure the continued growth and sustainability of your company? Or, to put it another way, *What's your play?*

Some Fortune 500 companies are doubling down on defensive strategies, promoting cost efficiency and productivity to protect their core business. Others are attempting to "future-proof" their business model through acquisitions. For instance, following a five-year acquisition spree, Walmart's September 2016 $3 billion deal with Jet.com signaled a serious commitment to competing with Amazon.com. General Motors Co. is clearly thinking about the future of car ownership with its (unsuccessful) attempted acquisition of Lyft.

None of this is wrong, and the logic makes sense. But is either approach sufficient to sustain the growth and health of your business and demark you from competition longer term? We believe not.

When challenged by a host of disruptors who are exploiting traditional market dynamics, finding growth opportunities in increasingly onerous regulatory and competitive conditions is hard these days. And harnessing the power of constantly evolving digital technology to break down well-established barriers to entry and devise new business models is a complex endeavor. To rise above the fray in this challenging context, you need to find ways to fight a battle you're well positioned to win. For leaders of big companies, this means capitalizing on an ability to do things the disruptors simply can't—set an ambitious vision, plan globally, invest strategically, and mobilize considerable resources to assert digital dominance. In other words, elevate above the level of the disruptors and transform your scale from a liability into an asset.

This may sound easy, but it isn't. Too many companies are still formulating their growth strategies based on traditional growth planning approaches—yearly cycles, historical analytics, and incremental thinking. With the velocity and uncertainty that characterize this new digital economy, traditional growth planning has reached the end of its useful shelf life. It just won't get you there.

Companies need to reframe the way strategy is formulated around three fundamental truths, and plot their next steps by embracing and owning these truths.

Truth 1: You can't analyze your way to the future; you need to invent it. Traditional analytical models that have been the bedrock of strategic planning for years are important as a means of establishing a foundational understanding of the world that exists today and the current opportunity areas. However, if you are looking to define a strategy that will enable your company to achieve disproportionate growth and create competitive advantage, you need to push beyond pure analysis.

What if the razor industry, dominated in the United States by giants Gillette and Schick, had looked beyond known competitors to anticipate the value in a direct-to-consumer subscription service? Would the e-commerce razor delivery company Dollar Shave Club have had such a meteoric rise? And would powerhouse Unilever, which acquired the startup in 2016, have expanded as meaningfully into the shaving business?

What's your play? Successful digital strategy requires a blend of deductive analysis and the type of inductive reasoning that powers the creative leaps that anticipate, and often open, fundamentally new markets. To help your organization see into the future, focus on creating an organizational culture that values a mix of inside-out and outside-in thinking.

Think about what the world will look like five to 10 years out—across a range of different industries, not just your own. In this future world, how do the needs of your current consumer change? Are there new opportunities outside of your existing consumer base or current product offerings?

Next, imagine removing your company's existing business and economic restraints from the equation. How would you use digital technology to overcome barriers and capture these opportunities? Run this exercise with a tiger team of out-of-the-box thinkers from across your organization. Let these inputs form the basis of your strategy formulation process.

Truth 2: Competitive evolution is no longer linear – it's exponential and disruptive. Your strategy needs to reflect these dynamics. Defining a destination point that's three to five years out is critical for focusing resources in a manner necessary for scale. However, the strategy can't just be a fixed gameplan that serves as marching orders for the foreseeable future. Strategy needs to take the form of a living, breathing process. Much like software development has evolved from traditional waterfall models to agile development, strategy formulation must embrace an approach that adjusts to rapidly shifting conditions in the market. Without dynamism and nimble capacity for adjustment, companies find themselves in situations where they simultaneously miss opportunities and create attractive windows for disruptors to attack.

Consider that as late as 2008, former Blockbuster CEO Jim Keyes stated that Netflix, the video-rental company that started out by mailing discs through the postal service and now streams media and video direct to consumers, was not "even on the radar screen in terms of competition." Yet within two years, Blockbuster, which built its name on brick-and-mortar video-rental shops, was bankrupt. Netflix, meanwhile, was on pace to acquire more than 65 million global streaming subscribers. The decision of Netflix CEO Reed Hastings to go all-in on streaming was a strategic bet against Blockbuster. Hastings knew that Blockbuster's traditional advantages—retail network, inventory, and sales staff—would quickly become unsustainable liabilities in the digital world.

What's your play? We've reached the end of the well-defined strategic planning cycle. You and your team must design a process through which you can continuously assess market dynamics, monitor the impact and opportunities presented by business and technology changes, and adjust direction to ensure you are constantly moving toward the goal. Don't make digital strategy a slave to your budgeting process. Create a dynamic series of sprints with a clear endgame in mind—and build flexible investment capacity to respond.

Truth 3: Ambition for growth isn't the problem. Your biggest hurdle may stem from an inability to catalyze the organization into action. Big companies don't lack ambition. More often than not, their primary challenge is getting leadership to put sufficient tension on the company to start the digital transition. Many organizations believe that defining the goal will be enough to spark the action required to capitalize on the opportunities. In reality, many of today's Fortune 500 companies are not designed to make bold moves. Multinational companies are optimized for efficiency, which makes it challenging for fundamentally new initiatives, ideas, and processes to gain traction. In keeping with the desire for operational efficiency, many companies are staffed with employees who are focused on performance excellence, implementation, and mitigating risk. Compounding these inherent challenges, capital flows to the lowest-risk ventures. Within public companies, this is exacerbated by the need to report quarterly earnings and manage earnings per share.

Barnes & Noble has grappled with these challenges for years—first failing to invest aggressively in online bookselling during the early rise of Amazon, and then playing the laggard in the e-reader race. Reeling from a series of brick-and-mortar closures over the last five years, the book retailer is struggling to galvanize action behind a solid growth strategy.

What's your play? From Six Sigma to agile development, strategies within today's modern organizations are devoted to constant process improvement. While there is value in these models, if your efforts are entirely focused on the perfection of an established process, it is difficult to devote time, energy, and investment to new digital plays. Provide space and acceptance across your teams for experimentation and ideation. Build time to explore new approaches into your organization's key performance indicators (KPIs) or objective and key results (OKRs).

And remember, for any growth plan to succeed, you must have a plan to catalyze action across the company. This isn't plain sailing, so how do you go about it? Give your team a common enemy to rally against. Create a persona of the competitor-of-the-future likely to disrupt your business—where will it happen, and how? Socialize this personification widely; make sure every department knows what kind of competitor the company is up against and why their department's input matters to win the battle.

Business leaders focused on generating sustainable growth are right to be concerned about potential disruption. But planning for the future by focusing exclusively on the realities of today is a shortsighted approach, and one that may result in the eventual demise of leading global companies. Big corporations can use their scale to fight back. To achieve the desired result, your digital growth strategy needs to be future focused, dynamic, and should provide a rallying cry that unites your entire business.

II

Building Data-Based Value

5

How to Monetize Your Data

Barbara H. Wixom and Jeanne W. Ross

The possession of rich amounts of data is hardly unique in today's world. Indeed, data itself is increasingly a commodity. But the ability to monetize data effectively—and not simply hoard it—can be a source of competitive advantage in the digital economy.

Companies can take three approaches to monetizing their data: (1) improving internal business processes and decisions, (2) wrapping information around core products and services, and (3) selling information offerings to new and existing markets. These approaches differ significantly in the types of capabilities and commitments they require, but each represents an important opportunity for a company to distinguish itself in the marketplace.

Theoretically, companies can pursue more than one approach to data monetization at the same time. In practice, adopting each approach requires management commitment to specific organizational changes and targeted technology and data management upgrades. Thus, it's best to identify your most promising opportunity and start there. In doing so, you will enhance your data in ways that will accelerate subsequent efforts related to the other approaches. More importantly, you'll build your company's capacity for monetizing its data.

Improving Internal Processes

Using data to improve operational processes and boost decision-making quality may not be the most glamorous path to monetizing data, but it is the most immediate. Executives often underestimate the financial returns that can be generated by using data to create operational efficiencies. Companies see positive results when they put data and analytics in the hands of employees who are positioned to make decisions, such as those who interact with customers, oversee product development, or run production processes. With data-based insights and clear decision rules, people can deliver more meaningful services, better assess and address customer demands, and optimize production.

When Satya Nadella became CEO of Microsoft Corp. in February 2014, he urged employees to find ways to improve the company's processes with data. Within sales, executives believed that, with the right tools and systems, they could improve the productivity of their salespeople by 30%. To do so, Microsoft's sales leaders sought to deploy tools that would help salespeople spend more of their time engaging with customers—and in more effective ways—by arming them with key computed insights, such as how likely a sale is to close and when.

To deliver actionable insights, sales executives first had to define shared concepts (for example, what is meant by "a lead"). They then needed to locate data sources that could be used to calculate performance. They quickly learned that sales data was located in too many different systems to easily create a comprehensive snapshot of a salesperson's business. Within a year, they created a new, integrated customer system that could produce 360-degree views of Microsoft's relationships with corporate customers, including what those customers bought, what issues they encountered, and how the company engaged with them.

The new system saved 10 to 15 minutes per sales opportunity by eliminating the need for Microsoft salespeople to manually search for and prepare data. The system also helped sales executives more accurately manage their pipelines; it used predictive analytics and machine learning to compute the likelihood of a successful sales engagement based on data that the salesperson provided about an opportunity. For example, buying and deploying enterprise software is complex and often requires a partner's involvement, so the system may calculate a higher likelihood for success when customers already have partners involved. Information about an opportunity's likelihood of success, along with suggestions on how to advance engagements along the sales pipeline, helped salespeople prioritize their leads and act in ways most likely to achieve their goals. Over time, Microsoft salespeople learned how to forecast more accurately (for example, the accuracy of forecasts regarding global accounts has risen from 55% to 70%), which has led to better sales-pipeline data and, in turn, improved pipeline management.

Wrapping Information around Products

Most companies have opportunities—often quite significant ones—to enrich their products, services, and customer experiences using data and analytics, a phenomenon that we call "wrapping." Companies are wrapping their offerings with data to escape commoditization and satisfy increasingly hard-to-please customers—with the goals of generating sales increases, higher prices, and deeper customer loyalty. FedEx Corp. was an early exemplar of wrapping when it introduced online package tracking as a free service in the 1990s. Now examples abound as companies bundle reporting, alerts, and other information to add value to products ranging from credit cards to health monitors.

Wrapping is a creative exercise in which companies identify what problems their customers have and then find ways to solve those problems using data and analytics. For example, Capital One Financial Corp., a diversified bank based in McLean, Virginia, learned that many of its credit card holders are concerned about fraudulent transactions but find the task of examining every charge to be tedious. So the company helps customers identify fraud more easily and quickly by displaying merchant logos and maps with each transaction in online statements. The visual cues jog cardholders' memories about whether they made a purchase or not. As a result, customers are more satisfied with the credit card and more likely to use it more often.

Johnson & Johnson has discovered the value of providing pattern identification to users of its health-monitoring products, including those for diabetics. The company offers its OneTouch Verio Sync Meter customers historical reporting on their blood glucose levels, along with tools to help them understand patterns of changes. The reporting is intended to help customers identify the possible causes for the glucose level variations and thus identify behavioral changes that can result in healthier living. Wrapping activities are best viewed as extensions of a company's product management processes. This means offering data and analytics to customers at the same level of quality as the core product. Doing so requires comparable levels of scrutiny and control. Most companies don't manage and cannot deliver data and analytics in this way. In fact, exposing data to customers could reveal quality problems and a lack of analytical sophistication. Thus, in most cases, wrapping requires companies to "up their game" in their information capabilities so that wrapping doesn't damage their reputation or undermine their value proposition. This effort may entail heavy investment in data-quality programs, advanced computing platforms (for instance, Hadoop), or data-science talent.

Selling Data

Many executives are eager to sell their company's data, convinced that it has inherent value and can generate important new revenues for the company. We caution that selling represents the hardest way to monetize data, mainly because it requires a unique business model that most companies are not set up to execute. Yet it can be done to potentially great effect under the right circumstances.

Based in Boston, Massachusetts, State Street Corp. is a financial services company that reported $10.6 billion in 2016 revenue. It provides products and services to institutional investors such as mutual funds, corporate and public retirement plans, and insurance companies. In 2013, State Street announced a new information-business division called State Street Global Exchange that would combine existing State Street data and analytics capabilities with new research to develop information-based solutions that clients would be willing to buy independently of the company's core services. State Street established a new division for the information business in recognition of its

Selling represents the hardest way to monetize data, mainly because it requires a unique business model that most companies are not set up to execute.

unique business model needs—something the company had not done in 30 years.

Even though it started out as a discrete unit, State Street Global Exchange focused on developing products that were tightly associated with State Street's core business. For example, State Street is one of the largest administrators of private equity assets, which means that it collects data about the financial capital that is not noted on a public exchange; this kind of data is of great value to markets that require an accurate representation of the private equity industry. State Street Global Exchange appreciated that the data was not automatically monetizable. Executives secured permission from 3,000 private equity clients to aggregate and anonymize that data—and then created an index that conveyed the financial performance of the private equity industry.

State Street leaders realized that they would need an entirely new operating model to support the information business. For one, sales processes had to change because, although State Street Global Exchange often sold to State Street clients, a buyer of Global Exchange products was frequently a different person or cost center than the kind of buyer traditional State Street products attract. In addition, the information business required salespeople with different selling experience and skills in selling stand-alone data and analytics-based products.

State Street understood that establishing an information business is hard and takes time. State Street Global Exchange had to learn to achieve balance between maintaining key ties with State Street (to create benefits from being a part of the larger organization) and responding quickly to new markets and new needs. Executives believe that State Street Global Exchange is gaining significant traction with its clients—and that their commitment will pay off. But we caution that such a model is not easy to

Only about a quarter of companies offer employees and customers easy access to the data they most need. You can't monetize data no one can use.

replicate. Other companies should think carefully about the operational capabilities, investment, and commitment required to successfully sell data.

The Importance of Accountability

Chances are you have two major obstacles to monetizing your data. The first is the accessibility and quality of your data. Our research has found that only about a quarter of companies offer employees and customers easy access to the data they most need. You can't monetize data no one can use.

The second obstacle is lack of accountability. All three approaches to data monetization require committed leaders who can redirect the behaviors of employees to deliver an important new value proposition.

Your inclination may be to solve the data quality issue first with big investments in new infrastructure. We propose that addressing the second issue of accountability will create urgency and commitment to addressing data quality issues—and so we recommend starting there.

Data monetization through process improvement requires strong process leaders. These leaders systematically use data to analyze the outcomes of existing processes and test hypotheses about proposed improvements. At Microsoft, for example, sales managers designated specific people to reshape and institutionalize new ways of selling. Process leaders are ultimately responsible for the design of best practices, the capture of the right data, the availability of tools, and the training of all staff regarding how to use data to do their jobs.

Data monetization through wrapping requires strong product leaders. These leaders treat the data that accompanies a core product or service much like any other product innovation—they hold it to the same quality standards. At Capital One, product leaders know the value of adding a data or analytics feature to a credit card because they predict—and then track—the lift in revenue from the information as well as the cost of providing it. Product leaders assemble teams to design experiments and methodologies that help analyze the impacts of information features and make appropriate adjustments.

Monetizing data by selling it requires a strong business-unit leader. That leader, in turn, must assemble a team that can launch and grow what is for most companies a new line of business. The head of that business will start by ensuring the value of the data and related services to potential customers. But the business head and his or her team must also design data, analytics, and dashboards to monitor the business and enable a rapid response to new business opportunities.

Each of the data-monetization strategies requires new processes, skills, and cultures to generate maximum returns. Companies with data-monetization experience have learned that it is insufficient to simply put data and tools into the hands of employees. Microsoft refined goals, cleaned up data, honed reports and algorithms, grew talent, and changed habits. Capital One and Johnson & Johnson reshaped product-management talent, platforms, and capabilities. State Street redesigned its organization and created a new profit formula that would generate stand-alone revenues from information.

Impressive results from data monetization do not transpire from single "aha" moments. Instead, they stem from a clear data-monetization strategy, combined with investment and commitment.

6

What's Your Data Worth?

James E. Short and Steve Todd

In 2016, Microsoft Corp. acquired the online professional network LinkedIn Corp. for $26.2 billion. Why did Microsoft consider LinkedIn to be so valuable? And how much of the price paid was for LinkedIn's user data—as opposed to its other assets? Globally, LinkedIn had 433 million registered users and approximately 100 million active users per month prior to the acquisition. Simple arithmetic tells us that Microsoft paid about $260 per monthly active user.

Did Microsoft pay a reasonable price for the LinkedIn user data? Microsoft must have thought so—and LinkedIn agreed. But the deal generated scrutiny from the rating agency Moody's Investors Service Inc., which conducted a review of Microsoft's credit rating after the deal was announced. What can be learned from the Microsoft-LinkedIn transaction about the valuation of user data? How can we determine if Microsoft—or any acquirer—paid a reasonable price?

The answers to these questions are not clear. But the subject is growing increasingly relevant as companies collect and analyze ever more data. Indeed, the multibillion-dollar deal between Microsoft and LinkedIn is just one recent example of data valuation coming to the fore. Another example occurred during the 2015 Chapter 11 bankruptcy proceedings of Caesars Entertainment Operating Corp. Inc., a subsidiary of the casino gaming company Caesars Entertainment Corp. One area of conflict was the data in Caesars' Total Rewards customer loyalty program; some creditors argued that the Total Rewards program data was worth $1 billion, making it, according to a *Wall Street Journal* article, "the most valuable asset in the bitter bankruptcy feud at Caesars Entertainment Corp." A 2016 report by a bankruptcy court examiner on the case noted instances where sold-off Caesars properties—having lost access to the customer analytics in the Total Rewards database—suffered a decline in earnings. But the report also observed that it might be difficult to sell the Total Rewards system to incorporate it into another company's loyalty program. Although the Total Rewards system was Caesars' most valuable asset, its value to an outside party was an open question.

As these examples illustrate, there is no formula for placing a precise price tag on data. But in both of these cases, there were parties who believed the data to be worth hundreds of millions of dollars.

Exploring Data Valuation

To research data valuation, we conducted interviews and collected secondary data on information activities in 36 companies and nonprofit organizations in North America and Europe. Most had annual revenues greater than $1 billion. They represented a wide range of industry sectors, including retail, health care, entertainment, manufacturing, transportation, and government.

Although our focus was on data value, we found that most of the organizations in our study were focused instead on the challenges of storing, protecting, accessing, and analyzing massive amounts of data—efforts for which the information technology (IT) function is primarily responsible. While the IT functions were highly effective in storing and protecting data, they alone cannot make the key decisions that transform data into business value. Our study lens, therefore, quickly expanded to include chief financial and marketing officers and, in the case of regulatory compliance, legal officers. Because the majority of the companies in our study did not have formal data valuation practices, we adjusted our methodology to focus on significant business events triggering the need for data valuation, such as mergers and acquisitions, bankruptcy filings, or acquisitions and sales of data assets. Rather than studying data value in the abstract, we looked at events that triggered the need for such valuation and that could be compared across organizations.

All the companies we studied were awash in data, and the volume of their stored data was growing on average by 40% per year. We expected this explosion of data would place pressure on management to know which data was most valuable. However, the majority of companies reported they had no formal data valuation policies in place. A few identified classification efforts that included value assessments. These efforts were time-consuming and complex. For example, one large financial group had a team working on a significant data classification effort that included the categories "critical," "important," and "other." Data was categorized as "other" when the value was judged to be context-specific. The team's goal was to classify hundreds of terabytes of data; after nine months, they had worked through fewer than 20.

The difficulty that this particular financial group encountered is typical. Valuing data can be complex and highly context-dependent. Value may be based on multiple attributes, including usage type and frequency, content, age, author, history, reputation, creation cost, revenue potential, security requirements, and legal importance. Data value may change over time in response to new priorities, litigation, or regulations. These factors are all relevant and difficult to quantify.

A Framework for Valuing Data

How, then, should companies formalize data valuation practices? Based on our research, we define data value as the composite of three sources of value: the asset, or stock, value; the activity value; and the expected, or future, value. Here's a breakdown of each value source:

Data as Strategic Asset

For most companies, monetizing data assets means looking at the value of customer data. This is not a new concept; the idea of monetizing customer data is as old as grocery store loyalty cards. Customer data can generate monetary value directly (when the data is sold, traded, or acquired) or indirectly (when a new product or service leveraging customer data is created, but the data itself is not sold). Companies can also combine publicly available and proprietary data to create unique data sets for sale or use.

How big is the market opportunity for data monetization? In a word: big. The Strategy& unit of PwC has estimated that, in the financial sector alone, the revenue from commercializing data will grow to $300 billion per year by 2018.

We define data value as the composite of three sources of value: (1) the asset, or stock, value; (2) the activity value; and (3) the expected, or future, value.

The Value of Data in Use

Data use is typically defined by the application—such as a customer relationship management system or general ledger—and frequency of use. The frequency of use is typically defined by the application workload, the transaction rate, and the frequency of data access.

The frequency of data usage brings up an interesting aspect of data value. Conventional, tangible assets generally exhibit decreasing returns to use. That is, they decrease in value the more they are used. But data has the potential—not always, but often—to increase in value the more it is used. That is, data viewed as an asset can exhibit increasing returns to use. For example, Google Inc.'s Waze navigation and traffic application integrates real-time crowdsourced data from drivers, so the Waze mapping data becomes more valuable as more people use it.

The major costs of data are in its capture, storage, and maintenance. The marginal costs of using it can be almost negligible. An additional factor is time of use: The right data at the right time—for example, transaction data collected during the Christmas retail sales season—may be of very high value.

Of course, usage-based definitions of value are two-sided; the value attached to each side of the activity is unlikely to be the same. For example, for a traveler lost in an unfamiliar city, mapping data sent to the traveler's cell phone may be of very high value for one use, but the traveler may never need that exact data again. On the other hand, the data provider may keep the data for other purposes—and use it over and over again—for a very long time.

Making implicit data policies explicit, codified, and sharable across the company is a first step in prioritizing data value.

The Expected Future Value of Data

Although the phrases "digital assets" or "data assets" are commonly used, there is no generally accepted definition of how these assets should be counted on balance sheets. In fact, if data assets are tracked and accounted for at all—a big "if"—they are typically commingled with other intangible assets, such as trademarks, patents, copyrights, and goodwill. There are a number of approaches to valuing intangible assets. For example, intangible assets can be valued on the basis of observable market-based transactions involving similar assets; on the income they produce or cash flow they generate through savings; or on the cost incurred to develop or replace them.

Making implicit data policies explicit, codified, and sharable across the company is a first step in prioritizing data value.

What Can Companies Do?

No matter which path a company chooses to embed data valuation into company-wide strategies, our research uncovered three practical steps that all companies can take.

Make valuation policies explicit and sharable across the company. It is critical to develop company-wide policies in this area. For example, is your company creating a data catalog so that all data assets are known? Are you tracking the usage of data assets, much like a company tracks the mileage on the cars or trucks it owns? Making implicit data policies explicit, codified, and sharable across the company is a first step in prioritizing data value.

A few companies in our sample were beginning to manually classify selected data sets by value. In one case, the triggering event was an internal security audit to assess data risk. In another, the triggering event was a desire to assess where in the organization the volume of data was growing rapidly and to examine closely the costs and value of that growth.

The strongest business case we found for data valuation was in the acquisition, sale, or divestiture of business units with significant data assets. We anticipate that in the future, some of the evolving responsibilities of chief data officers may include valuing company data for these purposes. But that role is too new for us to discern any aggregate trends at this time.

Build in-house data valuation expertise. Our study found that several companies were exploring ways to monetize data assets for sale or licensing to third parties. However, having data to sell is not the same thing as knowing how to sell it. Several of the companies relied on outside experts, rather than in-house expertise, to value their data. We anticipate this will change. Companies seeking to monetize their data assets will first need to address how to acquire and develop valuation expertise in their own organizations.

Decide whether top-down or bottom-up valuation processes are the most effective within the company. In the top-down approach to valuing data, companies identify their critical applications and assign a value to the data used in those applications, whether they are a mainframe transaction system, a customer relationship management system, or a product development system. Key steps include defining the main system linkages—that is, the systems that feed other systems—associating the data accessed by all linked systems, and measuring the data activity within the linked systems. This approach has the benefit of prioritizing where internal partnerships between IT and business units need to be built, if they are not already in place.

A second approach is to define data value heuristically—in effect, working up from a map of data usage across the core data sets in the company. Key steps in this approach include assessing data flows and linkages across data and applications, and producing a detailed analysis of data usage patterns. Companies may already have much of the required information in data storage devices and distributed systems.

Whichever approach is taken, the first step is to identify the business and technology events that trigger the business's need for valuation. A needs-based approach will help senior management prioritize and drive valuation strategies, moving the company forward in monetizing the current and future value of its digital assets.

The authors wish to acknowledge financial and research support from Dell EMC, Intel Corp., and Seagate Technology Inc. for this study; in addition, Cisco Systems Inc., IBM Corp., and NetApp Inc. provided financial and research support for earlier stages of the research. Several individuals made important contributions: Barry Rudolph of VelociData Inc., Douglas Laney at Gartner Inc., Barbara Latulippe and Bill Schmarzo at Dell EMC, and Terry Yoshii at Intel Corp.

7

Is Your Company Ready for HR Analytics?

Bart Baesens, Sophie De Winne, and Luc Sels

Big data and analytics are omnipresent in today's business environment. What's more, new technologies such as the Internet of things, the ever-expanding online social graph, and the emergence of open, public data only increase the need for deep analytical knowledge and skills. Many companies have already invested in big data and analytics to gain a better understanding of customer behavior. In fact, due to the introduction of various regulatory guidelines, some of the most mature analytical applications can be found in customer-focused areas in insurance, risk management, and financial fraud detection.

But what about leveraging big data and analytics to gain insights into another group of your company's key stakeholders: your employees? Although we see many companies ramping up investments in human resources (HR) analytics, we haven't seen many success stories in that area yet. Because HR analytics is "the new kid on the block" in business analytics applications, we believe its practitioners can substantially benefit from lessons learned in applying analytics to customer-focused areas—and thus avoid many rookie mistakes and expensive beginner traps.

Based upon our research and our consulting experience with customer-focused analytics, we offer four lessons about how to successfully leverage HR analytics to support your strategic workforce decisions. More specifically, we will juxtapose some of our recent research and industry insights from customer analytics against HR analytics and highlight four important spillovers.

Lesson 1: Model, measure, and manage your employee network dynamics. In our own research, we have found that ties between customers (such as social ties, credit card transactions made with the same merchants, or board membership ties between companies) are very meaningful in explaining and predicting collective behavior such as customer churn, customer response to marketing outreach, or fraud. It is our belief that these principles can be easily used to harvest some low-hanging fruit in HR analytics. In particular, a network can be constructed—with employees as the nodes and with the links between them based upon factors such as (anonymized) email exchanges, joint projects, colocation, and talent similarity, and possibly weighted for how recent such connections were. This network can then be leveraged to understand how smoothly new hires will blend into your workforce network; it also can be used to quantify the optimal mix, from a performance perspective, between behaviors that bring cohesiveness to the employee network and those that bring diversity.

By the same token, when laying off or firing employees, it is important to understand the social influence and impact of an employee in order to prevent viral effects or talent drain from happening to your network or company. Employees who serve as social influencers or community connectors within your organization's network should be carefully approached when making

What if your analytical model tells you that your hiring and firing policy is not at all sound – or is even discriminatory? That you are using the wrong selection criteria or are searching for the impossible?

firing decisions to avoid functionally disconnecting essential parts of your network.

Lesson 2: Big data and analytics are not magic. As with any new technology, it is important to set appropriate expectations from the outset. While they can be valuable tools, analytics techniques are not a panacea for all of your company's mission-critical and difficult HR decisions. After all, almost as soon as an analytical HR model is put into production, it becomes outdated, since its ecosystem (including but not limited to company strategy, the employee portfolio, and the macroeconomic environment) is constantly subject to change. Hence it is of key importance that the HR end user critically interprets, reflects, adjusts, and steers the outcomes of the analytical models using his or her business acumen, experience, and knowledge of the problem and organization. For example, what if your analytical model tells you that your hiring and firing policy is not at all sound—or is even discriminatory? That you are using the wrong selection criteria or are searching for the impossible? That the recent loss of customers can be traced back to the departure of a specific employee? Any unexpected yet valid analytical findings should be approached in a careful and thoughtful way. Obviously, this requires HR managers with a mindset that is both informed and open.

Lesson 3: Analytical HR models should do more than provide statistical performance – they should provide business insights. A typical rookie mistake when deploying analytical models in any business context is a blind obsession with statistical performance (such as fit, correlation, R-squared, etc.) and overly complex analytical models. Statistical performance is important, but analytical HR models should do more. Two other important performance criteria are model interpretability and compliance.

Interpretability means that any HR decision based upon analytics should be properly motivated and can be simply explained to all stakeholders involved. This quest for simplicity discourages the use of overly complex analytical models that focus more on statistical performance than on proper business insight.

Another key performance criterion concerns model compliance. Safeguarding regulations, privacy, and ethical responsibilities is crucial to successfully deploying HR analytics. This is especially important in HR applications. Analytical models should always be interpreted with caution, and gender equality and diversity should be respected when selecting the data to build your analytical HR models.

Lesson 4: Backtest the impact of your analytical workforce models. In customer analytics, the average lifespan of a model is two to three years, and we have no reason to believe that this will be different in HR analytics. However, given the impact of HR decisions on the organization and on individuals, it is important that analytical models in HR are constantly backtested by contrasting the predictions against reality, so that any degradation in performance can be immediately noticed and acted upon. For example, from a hiring perspective, both the prehire effectiveness (which recruitment channels give us the candidates with the right profile?) and posthire effectiveness (which recruitment channels gave us the best candidates?) should be constantly evaluated.

We believe the time is right to boost your investments in HR analytics. And once your HR analytics efforts have matured, we look forward to the next transformative step for organizations.

That, we think, will take place when organizations can bring together findings from HR analytics with those from customer analytics. Then companies can more fully understand the relationships between their two key sets of human assets: employees and customers.

8

Why Your Company Needs Data Translators

Chris Brady, Mike Forde, and Simon Chadwick

Over the past two years, we've worked extensively with leaders in the world of professional sports, a field known for its use of analytics. An emergent theme of our work has been the persistent cultural divide between the decision makers on the field and the data analysts who crunch numbers off of it.

Our work has included a series of research workshops to discuss transatlantic and cross-sector issues around performance management in professional sports. A key issue that emerged from these meetings was the recognition of this consistent disconnect within performance management practice between big data analysts and the decision makers they support. This is evidenced by the predominantly dismissive attitude of many executive decision makers (general managers, head coaches, CEOs, COOs, etc.) to both the data itself and those responsible for delivering it—an attitude often born largely out of ignorance or fear. The research group believed that bridging this cultural gap would provide considerable competitive advantage to any organization concerned with high performance.

What's more, this issue transcends the world of professional sports. Whatever your industry, it's likely that misunderstandings between quants and frontline decision makers are a challenge your business is confronting, too. As Jeanne G. Harris and Vijay Mehrotra noted in a 2014 article in *MIT Sloan Management Review*, the problem is one of communication. "A common complaint is that data scientists are aloof and seem uninterested in the professional lives and business problems of less-technical coworkers," they wrote. "They don't see a need to explain or talk about the implications of their insights, which makes it difficult for them to partner effectively with professionals whose business expertise lies outside of the technical realm."

What is to be done? From our work with successful sports leaders, we accept that there is a significant gap between the quants and the decision makers, a gap that we call the "interpretation gap." We believe that those who are needed to fill that gap are what we call "data translators." While some have argued that data scientists can bridge the gap, we think that, in many cases, the data translator role can best be filled by domain experts. To date, many businesses have been trying to bridge the gap by teaching the quants (often recent graduates) about the business in which they operate. But in some cases, it may be easier for domain experts, with deep knowledge of the business in which they are engaged and the requisite interpersonal skills, to obtain sufficient knowledge about data analysis to act as the translator for data scientists than for data scientists to gain enough knowledge about the domain, especially the language of that domain. Domain expertise requires a high level of practical experience, which is difficult to acquire on a theoretical basis, and it also lends itself more readily to the storytelling ability that must be an essential skill of the data translators.

Here are some of the issues we think companies need data translators' help to address.

Data Hubris

Translating analytics into a language decision makers understand is not as simple as it sounds. Among other things, the person doing the translating—whether it's a quant or a data translator serving as a liaison between the quant and an executive decision maker—needs to avoid what's referred to as data hubris. In a 2014 *Science* article about the potential pitfalls of relying on big data, David Lazer and his coauthors described data hubris as "the often implicit assumption that big data are a substitute for, rather than a supplement to, traditional data collection and analysis."

In the sports world, the mistake of data hubris is commonplace. A quant analyzes statistics and draws firm conclusions about individual players, to the point where the quant believes the numbers, in a vacuum, provide a clearer picture than what the coach observes every day with his own eyes—in practice, in games, and in the locker room.

At the heart of this conflict is a false dichotomy between numbers and intuition. In reality, decision makers must seek what R. C. Buford, general manager of the San Antonio Spurs basketball team, described, in an interview with us, as "alignment of the multivariables—the eyes, the ears, the numbers." In other words, organizations should use analytics and firsthand observations in a complementary way to form a holistic opinion, rather than lean too heavily on only data or only observations.

Decision-Making Biases

Whether you're a quant or a decision maker who balances both observations and numbers, you must remain aware that any point of view, even one derived from extensive research and rock-solid facts, carries potential biases.

For example, one bias that dampens the utility of data-driven intelligence is commonly referred to as overconfidence bias—when an individual's confidence in his or her own judgment is at odds with reality. Of course, the individual may have perfectly good reasons to be confident, as opposed to overconfident. Perhaps he or she has a stellar track record or is taking a position based on thorough research. But that doesn't mean he or she can't be wrong, especially if the topic is one for which making predictions is an inherently tricky business.

In sports, one of these unpredictable topics is talent evaluation. How can teams assess which up-and-coming young athletes will perform best as pros? Teams invest heavily in scouting and player evaluation, but they still make mistakes, because predicting individual performance is far from an exact science.

And yet, precisely because teams invest so heavily in evaluation, they can often be overconfident. "Even the smartest guys in the world, the guys who spend hours with game film, can't predict [the subsequent performance of football draft choices] with much success," Cade Massey, a professor at the University of Pennsylvania's Wharton School who has studied the National Football League (NFL) draft pick, once told *The New York Times*. "There is no crime in that. The crime is thinking you can predict it."

Another issue to be careful about is emotional bias. In an interview with us, Billy Beane, executive vice president of baseball operations for the Oakland Athletics baseball team, described emotional bias as the consequence of being a decision maker in the public eye, constantly second-guessed by fans, customers, and social media followers. Emotional bias occurs when the decision maker lets the outside noise influence his decisions. "All decisions are now public decisions; everyone is an expert," Beane told us. "There is permanent media scrutiny, and it must have some sort of effect on decision making. The decision maker needs to eliminate the noise."

Need for Linguistic Common Ground

Another powerful theme that emerged from our research is the significance of the communication barrier. It is apparent to us that leaders in senior management do not speak the same language as the analysts.

We learned that decision makers are seeking clearer ways to receive complex insights. They want analysts to speak to them in plain language, abetted by visuals, so they can easily absorb the meaning of the data. Our findings are consistent with a recent IBM survey suggesting that executives intend to replace standard reporting techniques with approaches that bring otherwise dry information to life. These approaches include data visualization, process simulation, text and voice analytics, and social media analysis.

The Importance of Translation

Sig Mejdal, special assistant to the general manager, process improvement, for the Houston Astros baseball team, has pointed out that most decision makers are "not conversant with the scientific method. So we have to change our language." By "we," Mejdal means quants like himself. To bridge the gap that Mejdal describes, we suggest finding the people in your organization capable of conversing with both the quants and the decision makers. We call these talented communicators "translators," since in a manner of speaking, they are abetting understanding between two different cultures.

The key to effective translation is understanding each of the figurative languages, as well as each of the figurative cultures. For example, Del Harris, a well-known coach in the National Basketball Association (NBA), has been an effective translator during his career, helping the coaching staff make sense of the numbers and helping the quants make sense of the coaching staff. At the 2015 MIT Sloan Sports Analytics Conference, he explained how, at one team where he was an assistant coach, the analytics came directly to him, as opposed to the head coach, because without his ability to translate, the head coach "couldn't care less about that sort of thing."

There's more to effective translation than simply rendering scientific language in plain terms. The best translators also frame the information in a way those receiving the translation will find useful. In the plainest language, a translator must ask one blunt question: How does this data help the person I'm speaking to?

The Skills Translators Need

From our experiences, we have created a checklist of skills that we believe the best data translators will possess:

- Sufficient knowledge of the business to pass the "street cred" test with executive decision makers
- Sufficient analytics knowledge—or a willingness and ability to acquire it—to communicate effectively with the organization's data scientists
- The confidence to speak the truth to executives, peers, and subordinates

- A willingness to search for deeper knowledge about everything;
- The drive to create both questions and answers in a form others find accessible
- An extremely high sense of quality standards and attention to detail
- The ability to engage in team or organizational meetings without being asked for input

And remember: It's also possible to teach translator skills to the talent you already have in-house. You can do this by reinforcing two important communication habits:

Connect with decision makers through questions, not assertions. Especially with skeptical decision makers, it's essential not to be overly assertive at the outset. Teach your quants to ask questions that enable the decision makers to come up with the answer, ostensibly by themselves.

Create analogies around anecdotes that resonate with the decision makers. These could be the stories of successful analytic interventions. In sports, these stories could carry themes such as "This athlete wasn't expected to make it this far—but he has." Or: "This strategy was counterintuitive, but it worked. Here's why."

Bridging the cultural gap between domain specialists and analytics specialists within organizations with an interpretation function performed by a data translator can begin to address the disparity between the claims for big data and its reality. That process begins with recognizing the limitations of what numbers and intuition can do separately.

III

Operational Makeovers

9

Sales Gets a Machine-Learning Makeover

H. James Wilson, Narendra Mulani, and Allan Alter

We live in a data-saturated world where a great many of our interactions with other humans happen online. It makes sense then that one of the most human of business activities—sales—is currently undergoing a digital renaissance. While the sales function has historically relied on metrics, today there is far more sales-centric data, and far richer data, than ever. It comes from social media, from website interactions, and from A/B tests, just to name a few.

To help make sense of all the available data and to improve sales effectiveness and efficiency, organizations are turning to machine learning. Smart machines are becoming trusted sidekicks in sales departments as they make opaque processes more transparent, provide analysis to inform decision making, and offload low-value tasks.

In our survey of executives at 168 large companies with at least $500 million in annual revenue, 76% of respondents said they are targeting higher sales growth with machine learning, the kind of artificial intelligence software that continuously learns from big data and optimizes recommendations in real time to sales staff. Moreover, more than two out of five companies have already implemented machine learning in sales and marketing.

Our research shows large companies are applying machine learning to sales processes along three dimensions, each of which adds algorithmic rigor to human intelligence and intuition, creating a dynamic new formula they hope will boost sales. The first dimension allows for a scientific approach—with data and clarity of process—in sales interactions. The second enables more data-driven experimentation within a sales and marketing environment. The third uses science to create more time to sell, that is, by automating the administrative tasks that get in the way of managing accounts, finding leads, and closing deals. Taken across these three dimensions, machine learning is creating speedier, more scientific processes for generating sales revenue.

Before machine learning came along, static databases, analytics from historical data, and experience and instinct steered execution—with performance improvements coming in set increments over time. With machine learning, real-time data can drive actions and process change along a continuous path. Hypotheses can be quickly formed, tested, and revised, enabling a new kind of workflow that can dramatically outperform previous ones. In our survey, 38% of respondents credited machine learning for improvements in their key performance indicators for sales—such as new leads, upsells, and sales cycle times—by a factor of 2 or more, while another 41% created improvements by a factor of 5 or more.

Bringing Science to Analyzing Social Cues

Historically in sales, a field rep might meet with a potential customer face-to-face and read nonverbal cues like nodding or frowning to determine his or her next move. But in a digital world—without the benefit of physical social cues—selling becomes an opaque process that can be difficult to deconstruct. If a prospect doesn't work out, it can be hard to find the errors that could be corrected in future attempts.

But what if a salesperson could know with confidence when a potential customer is ready to buy? One company called 6sense offers a product that provides digital predictive buying signals to help sales professionals pinpoint the optimal time to approach prospects. By analyzing the online behavior of potential customers who visit a client's site—as well as third-party data from a variety of publicly available sources, including social media—6sense provides a better picture of interest and if and when a potential customer might be ready to buy.

The company analyzes website data on a large scale, using machine learning to fine-tune predictions. With the right data in hand, sales teams can identify prospects more quickly, while targeting sales pitches at the right time, with a higher likelihood of success. With more data about potential customers, sales professionals have the ability to test different approaches, spending more time fine-tuning pitches rather than chasing false leads.

Data-Driven Sales Experiments

Machine learning can also enable more effective A/B website testing, eliminating the bottlenecks often associated with sales experiments. Fewer bottlenecks means more speed: One-third of our survey respondents say they have accelerated sales processes by double or more, while another one-third claim increases of five times or more. Adobe Target is one software tool that allows nontechnical salespeople and marketers to quickly modify websites to deploy large numbers of A/B tests. Based on data from website interactions, machine-learning algorithms find and suggest the best content to tweak, as well as help sales and marketing staff validate assumptions after developing a test.

A startup called Optimizely uses machine learning to run A/B tests on pricing strategy. In an experiment with marketing firm Bizible, Optimizely integrated its experimentation software with Salesforce. The result was a dashboard that displays experimental variables—the original prices and the test prices—as well as information on lead, contact, case, etc. Software ensured that pricing was consistent across a range of IP addresses so that potential customers at the same companies would see the same prices. The test only ran 30 days, but the results were compelling. New higher prices led to fewer opportunities, but those opportunities brought in 25% higher value on average.

When intelligent automation is used to facilitate science within organizations, it provides opportunities to test new actions and processes that can lead to revenue growth. Machine learning can act like an assistant at the lab bench, logging data and suggesting new experimental approaches. It can illuminate previously opaque processes and free salespeople and marketers to build their own experiments with clarity and confidence.

Automated Science for Sales Efficiency

Another important aspect of machine learning is that it can optimize processes behind the scenes. More than nine out of 10 companies in our survey said they agreed that machine learning is improving processes in real time without human intervention.

Algorithms can conduct automated science experiments with data as it arrives without any human intervention. In the case of sales, machine learning can minimize time spent on administrative tasks and eliminate steps that take time away from interactions with customers. The end result can be a significant reduction in sales cycle times.

Historically, many sales and marketing teams have attempted to increase efficiency with one-off tricks that prove difficult if not impossible to scale—writing their own macros or personalizing spreadsheets, for instance. Conversely, machine learning algorithms that automate administrative tasks or provide just-in-time customer predictions can be easily standardized and implemented across teams.

Gainsight, a company that offers software to manage sales and customer service more effectively, helped the online questionnaire service SurveyMonkey create automated alerts to ensure that all team members were up to date on renewals, invoicing, and upsell opportunities. Using Gainsight's technology, SurveyMonkey cut the process time to send an invoice by about a third.

Another company called Anaplan is helping Hewlett-Packard reduce the time spent gathering sales data from a month to three days, effectively a 10-fold improvement. Instead of churning through month-old data, sales teams can make decisions with fresh, up-to-date analysis, allowing sales staff to spend time on higher-value tasks. Similarly, a machine-learning company called Aviso, working with an enterprise cloud company called Nutanix, can compress a 12-hour task of compiling sales reports into four minutes.

Whether machine learning facilitates analysis, experimentation, or automation, it provides real value for sales and marketing teams. In some cases, salespeople and marketers gain confidence and clarity in processes that were previously opaque, enabling a more systematic and consistent approach to client interaction. In others, machine learning runs the experiments behind the scenes by pruning processes and allowing salespeople to attend to higher-value tasks. While we remain in the early stages of bringing the full value of machine learning to bear in sales (and elsewhere in the organization), what's clear already is that machine learning holds the potential to find significant hidden revenue where there were previously only marginal gains.

10

A New Approach to Automating Services

Mary C. Lacity and Leslie P. Willcocks

For more than 130 years, managers have, in effect, been trying to get humans to act like robots by structuring, routinizing, and measuring work—all under the guise of organizational efficiency.[1] The automation software that is being developed today[2] enables a reversal of this process. We are now able to use software robots to amplify and augment distinctive human strengths, enabling large economic gains and more satisfying work. However, given the widespread skepticism and fears about how many types of employment will fare in the future, managers are in a difficult spot. Media headlines such as the "Rise of the Robots: Technology and the Threat of a Jobless Future"[3] and "A World Without Work"[4] only serve to fuel the anxiety.

Although the term "robot" brings to mind visions of electromechanical machines that perform human tasks, the term as it relates to service automation refers to something less threatening: software that performs certain repetitive and dreary service tasks previously performed by humans, so that humans can focus on more unstructured and interesting tasks. Service automation includes a variety of tools and platforms that have various capabilities. While conducting research for this article, we interviewed people who used a variety of terms to discuss service automation (see "About the Research"). To help make sense of the landscape, we classified the tools along a service automation continuum based on the specific types of data and processes.

This article focuses on what we call *robotic process automation* —software tools and platforms that can automate rules-based processes that involve structured data and deterministic outcomes. The great majority of the 16 cases we researched involved robotic process automation. We focus on this area (as opposed to a more advanced automation technology known as cognitive automation) because this is where most companies today begin their service automation journeys.[5]

How do companies apply robotic process automation? A broad range of service tasks are suitable for such automation. Companies we studied used robotic process automation for tasks including those associated with validating the sale of insurance premiums, generating utility bills, paying health care insurance claims, keeping employee records up-to-date, and even generating news stories. Consider the example of Xchanging PLC,[6] a London-based business process and technology services provider that has clients across a variety of industry sectors. For one of its clients in the insurance sector, Xchanging processes insurance premiums so insurance brokers get paid. When brokers sell an insurance policy, they submit notices using a variety of inputs (email, fax, spreadsheets, etc.) to Xchanging, which manages the multistep process of validating the sale.

Previously, Xchanging's human operators managed the transactions manually. They organized the data, checked it for completeness and accuracy, worked with the insurance brokers to correct errors, extracted other necessary data from online sources, and then created and posted the official sales records.

Service Characteristics

	Realm of robotic process automation	Realm of cognitive automation
Data	• Structured	• Unstructured
Processes	• Rules-based	• Inference-based
Outcomes	• Single correct answer	• Set of likely answers
	Robotic process automation tools are designed to be used by subject matter experts to automate tasks that use rules to process structured data, resulting in a single correct answer—in other words, ***a deterministic outcome.***	*Cognitive automation tools are designed to be used by IT experts to automate tasks that use inferences to interpret unstructured data, resulting in a set of likely answers, as opposed to a single answer—in other words,* ***a probabilistic outcome.***

The Service Automation Landscape

The plethora of software tools and terms to describe software designed to automate services can be very confusing. To help make sense of the service automation landscape, we suggest avoiding the jargon and instead focusing on the service characteristics that the tools are designed to help automate. We consider two broad classes of service automation tools: robotic process automation and cognitive automation. Each class of tools is designed to deal with specific types of data and processes.

Humans still handle the unstructured parts of the work, such as formatting the inputs into structured data, passing the data to the software robots, and interacting with insurance brokers. However, the structured parts of the process, including finding the errors, retrieving the online data, creating the official sales record, and notifying brokers when the process is complete, is managed by the robotic process automation software.

Whereas it used to take a team of humans several days to complete 500 notices, today a properly trained software robot working with the help of a few humans can do the same amount of work in around 30 minutes. The software can be scaled up and down to meet changing workloads. Beyond this particular process, Xchanging has developed an enterprise-wide service automation capability in other areas as well, which it has deployed on clients' processes as well as on its own. By early 2016, Xchanging had automated 14 core processes and deployed 27 software robots; collectively, these robots were processing 120,000 transactions per month, with cost savings averaging 30% per process.

Xchanging isn't alone in experiencing benefits from robotic process automation. Similar gains were reported by other organizations we studied. Typically, companies indicated that they saw returns on investment of 30% or more during the first year of robotic process automation implementation;[7] however, because of the nature of our study of early adopters, we can't say whether or not such returns on investment are typical.

About the Research

We conducted empirical research on service automation to answer three questions: (1) Why are companies adopting service automation? (2) What outcomes are they achieving? and (3) What practices distinguish service automation outcomes? To answer these questions, we conducted two surveys of professionals attending the International Association of Outsourcing Professionals world summits in 2015 and 2016 and conducted interviews with 48 people, including service automation adopters, software providers, and management consultants across the major business sectors.

In the course of our research, we collected 16 service automation adoption stories: 14 companies adopted robotic process automation, and two adopted cognitive automation tools. Depending on the subjects' availability and preferences, we conducted interviews in person, over the phone, and through email. We posed a number of questions pertaining to their service automation adoption, the business value delivered, and lessons learned. We also interviewed software provider representatives to discuss their companies' automation capabilities, challenges they help their clients overcome, and the future they envisioned for service automation. We asked advisors questions pertaining to client service automation adoption, effects on outsourcing, automation tool capabilities, and the future of work as a consequence of automation. Of the 16 research sites we focused on, seven companies were headquartered in the United Kingdom, five in the United States, and one each in Germany, France, the Netherlands, and Russia. The organizations represented 11 industries, including health care, energy, telecommunications, media, financial and accounting services, and transportation.

This research was conducted with support and funding from the Outsourcing Unit at the London School of Economics and Political Science; Information Systems Group, a technology advisory services company based in Stamford, Connecticut; and Blue Prism Group PLC, a UK–based robotic process automation software company. Blue Prism introduced us to 10 of the companies we used for case studies; we conducted our interviews at those companies independently. The remaining company case studies were also developed from interviews we conducted independently.

Beyond the financial benefits, the automation solutions improved service speed and quality, expanded service availability to 24 hours, and increased regulatory compliance. Software robots executed structured tasks precisely and quickly—and did so without the need to eat or sleep. When the software robots were partnered with humans, the combined human-robot teams were high-performing. Moreover, they easily scaled to take on a higher volume of structured work when needed, with humans filling in the gaps that required on-the-fly problem-solving and hands-on customer care.

By studying organizations that were early adopters of software robots, we saw how companies could generate tangible benefits via service innovations. They achieved benefits in three ways: (1) by developing an approach to service automation supported by top management, (2) by initiating effective processes that deliver value to customers and employees, and (3) by building enterprise-wide skills and capabilities. Managers interested in capturing the benefits of service automation need to pursue all three avenues.

Developing a Service Automation Strategy

Companies that captured the full benefits of service automation took a long-term view. Whereas some companies approach service automation as a way to achieve quick wins for the business, we found that those that undertook it as part of a broader and more integrated business strategy were able to achieve more substantial gains.

Service automation enables a broader business enterprise strategy. Our experience indicated that the businesses with the best outcomes didn't have a service automation strategy per se; instead, they had strategies that defined the organization's long-term goals, such as creating a more flexible workforce or expanding services without expanding head count. These strategies were driven by management and enabled, in part, by service automation; it was a key component of the business transformation.

The Associated Press, a New York City–based news cooperative, offers a good example. In 2014, the AP began offering its newspapers and other media organizations automated corporate earnings reports. The AP was eager to find ways to expand its news coverage without increasing costs and to enhance its brand. Lou Ferrara, then an AP vice president and managing editor, spearheaded a service automation initiative. He found that reporters preferred to cover stories that required creativity and that this was how they added the most value. Most reporters disliked assignments that were highly structured, such as reporting on corporate earnings. By automating the corporate earnings reports, the AP was able to expand its coverage at no additional cost. In fact, the volume of its earnings reports rose from 300 reports per quarter when humans wrote them to more than 3,700 reports with a software robot. In addition to producing more content, the automation freed up time for the equivalent of three full-time reporters. The company's unionized journalists kept their jobs, and clients were happy with the quality and the quick delivery. After the AP introduced the automated corporate reports, it initiated a similar automation project to expand coverage of college sports news.

Strategic service automation requires support from senior management. Organizations where the C-suite supported and promoted service automation tended to achieve more strategic benefit from service automation than those where the support was at the divisional or information technology (IT) level. Without support from the top, there isn't sufficient breadth of influence or application, and people from other parts of the organization may treat the robotic process automation project as a curiosity.

The experience of a major European gas and electric utility that we studied highlights this point. Eager to improve its service and control its operating costs in order to reduce the need for rate increases, the utility's senior management, led by the company's CEO, embraced automation beginning in 2008. One nagging issue the utility had grappled with was how to verify the meter readings household residents submitted on paper rate cards or by phone, or that came via independent meter readers.

After meter readings were submitted, this information was digitized and entered into a system that asked if the information made sense. Did the reading fall within normal energy consumption ranges? Was there anything incongruous about it (such as a user adding electricity to the grid rather than consuming it)? The outliers were spit out as exceptions and sent to humans for verification. Some of the readings were easy to sort out; others required calling customers. With robotic process automation, only the truly unusual cases required human intervention. In the division where automation was first implemented, the utility was able to reduce the number of humans assigned to this activity from 30 to about 12.[8] In addition to payroll savings, the organization was able to improve quality, consistency, and speed of problem resolutions. By early 2016, the utility was deploying hundreds of software robots, which allowed it to automate about 25% of its back-office work on meter management, customer

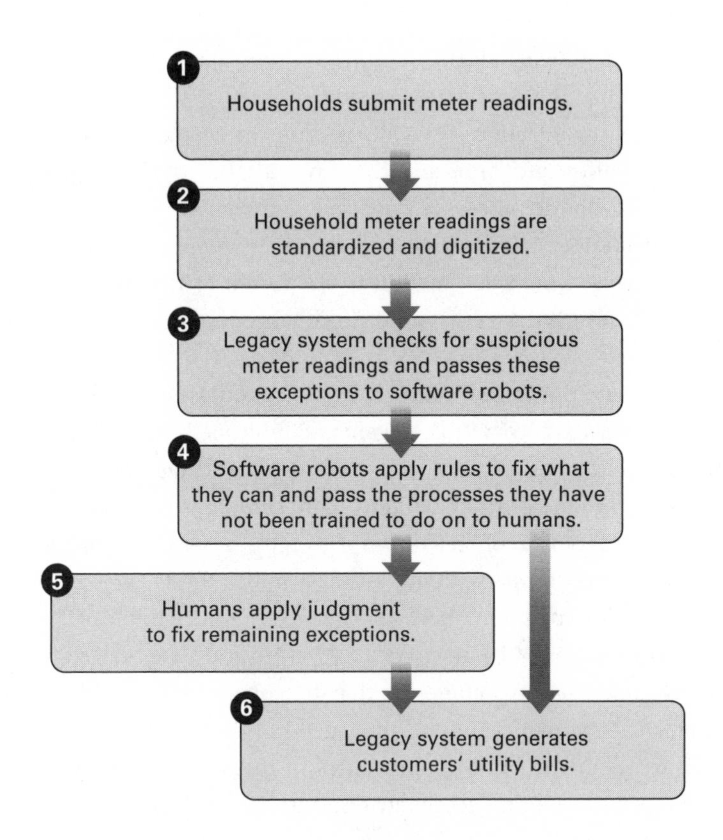

How Humans and Software Robots Work Together at a Utility Company

The European utility company we studied verifies household meter readings before generating a customer's utility bill. After robotic automation was applied to the process, the software robots could handle enough exceptions to free up 60% of the humans from this task; those who have continued working in this area work on only the most unusual exceptions.

billing, account management, consumption management, segmentation, and exception processing.

In this organization, the CEO became an evangelist for the transformation programs and the ways in which technologies, including robotic process automation, contributed to them. He gave regular pep talks to divisional managers about the strategic importance of such automation to the future of the company, and that has played a critical role internally.

Service automation can deliver multiple benefits. Organizations can use service automation to generate multiple business benefits, including cost savings, improved customer experience, and, as we saw in the case of Xchanging, higher employee satisfaction.

The experience of Telefónica UK Ltd., a telecommunications services company doing business under the O2 brand in the United Kingdom, offers another excellent example. (The company is owned by Telefónica S.A., a Madrid-based telecommunications services company that has operations in Europe, Asia, and the Americas.) Some managers at O2 expected automation to result in major opportunities to reduce the company's employee head count, accelerate response time to customer queries and activation of phone services, and reduce the number of customer calls inquiring about service status. The company began in 2010 by automating the structured tasks associated with two processes: the process that updated digital records to reassign a customer's phone number from his or her old phone to a new phone; and the process used for applying precalculated credit amounts to verify that a customer had sufficient credit to permit orders to be processed in advance of payment.

In the space of five years, O2 had automated nearly 35% of its back-office services. In 2015, the company's software robots were processing between 400,000 and 500,000 transactions each month. For some customer-facing processes (for example, phone activation), turnaround times that previously were measured in days were cut to just minutes. What's more, the service automation enabled greater workforce flexibility. To support a new product launch, for example, the "robotic" workforce could be doubled almost instantly and then scaled back after the initial market surge.

Organizations seeking to automate services have multiple sourcing options. A peculiarity of our research sample was that all of the organizations we examined adopted service automation themselves and relied on the help of a service automation tool provider to get started.[9] For example, when the European gas and electric company adopted robotic process automation, its tool provider trained a handful of client employees and provided mentoring, consulting, and codevelopment for the first set of automated processes. Initially, about 80% of the robotic process automation team was from the tool provider's staff, with the rest from the utility's staff. As the utility gained experience over the next nine months and automated additional processes, the ratio of outside staff to inside staff flipped. However, our survey data (along with our prior research on business-process outsourcing[10]) has led us to think it's important for organizations that are considering robotic process automation and other service automation technologies to evaluate a broad spectrum of sourcing options to determine what meets their needs best. The options include:

- *Insourcing:* buying service automation software licenses directly from a service automation provider;

- *Insourcing and consulting:* buying licenses directly from a service automation provider and engaging a consulting firm for services and configuration;
- *Outsourcing with a traditional business-process outsourcing (BPO) provider:* buying service automation as part of a suite of integrated services delivered by a traditional BPO provider;
- *Outsourcing with a new provider:* buying service automation from a new outsourcing provider that specializes in service automation;
- *Cloud sourcing:* buying service automation as a cloud service.[11]

In our survey, we found that insourcing enabled client organizations to achieve high levels of control and allowed them to keep whatever cost savings they generated. However, other options offer benefits as well. For example, many traditional business-process outsourcing providers have developed significant automation capabilities. The benefit of engaging an experienced service provider is that such a provider often has a suite of integrated services that combines low-cost offshore labor, process excellence, experience in managing change, and technology expertise. There are newer companies that also specialize in service automation. Whereas traditional business-process outsourcing providers integrate automation into their overall service delivery, the new players are focused on helping customers learn about and apply the new breed of robotic process automation tools. The most promising option involves placing software robots in the cloud, where they can be copied and deployed across the network. Indeed, if it can take months to train a software robot to master a complex task, it might only take a few minutes (or even seconds) to transfer its capabilities to another software robot in the cloud.

Initiating Effective Automation Processes

Once executives have developed their strategies, they must enable execution. They need to have committed middle managers who can help deliver the service automation vision. To ensure that the best processes to automate are identified, it's also important that business operations, rather than the IT department, leads the service automation initiative. However, organizations should involve IT professionals early to avoid risks to the organization, such as exposing customer data. Along the way, companies need to pay close attention to internal communications to inform employees about the service automation strategy and timing and its effects on employees.

Identify sponsors, program champions, and program managers. A successful robotic process automation project requires multiple levels of management support. To begin with, projects need sponsors—people who initiate the idea, underwrite the resources, and push for the technology's adoption and use. Depending on the company's ambitions, the sponsor might be part of the C-suite or a middle manager in charge of a department such as shared services.

Whereas a sponsor might only spend 2% to 3% of his or her time on a project, program champions take a more hands-on role, spending anywhere from 40% to 80% of their time communicating the vision, maintaining motivation among team members, and interacting with stakeholders, including senior management. In addition, projects need strong program managers who know how to get the projects delivered within budget and on schedule. At Xchanging, the project sponsor was the CEO of the company's insurance business. The project champion, for his part, had lots of experience in leading lean process projects; he served as both project champion and program manager.

Allow business operations to take the lead. People preparing to embark on service automation projects often ask, "Where should service automation originate—in business operations, IT, or outsourcing provider companies?" The companies we featured above—Xchanging, AP, Telefónica UK (O2), and the utility—initiated their service automation projects in business operations. Since we were studying the automation of business processes (not the automation of IT processes), it makes sense that business operations should lead this kind of service automation. In fact, several of the people we interviewed were adamant that business operations needs to own these automation programs.

Business operations tends to be in the best position to select tasks within their processes that are most suitable for automation. Managers in these areas know which processes fit the minimum criteria for automation: The managers can identify tasks that use structured data, have explicit and well-documented rules, churn out high transaction volumes, and are stable.[12] As we saw with the European utility's and Xchanging's work in the insurance industry, business operations selected structured tasks associated with end-to-end processes for automation and left the tasks requiring judgment and social interaction for humans. Business operations groups are also in the best position to prioritize automation projects that will yield the best outcomes for customers and employees.

Pinpoint what you're trying to achieve and how it will be perceived by customers and/or employees. Many new technologies overpromise and underdeliver. Before you embark on a service automation project, make sure stakeholders such as customers and employees are attracted to the supposed benefits. In the case of the AP's automated earnings reports, customers liked the idea of expanded corporate earnings coverage, and journalists were positive about the reframing of their job responsibilities.[13]

VHA Inc., a health care network of not-for-profit hospitals based in Irving, Texas, provides services such as centralized procurement to its affiliate hospitals, which results in lower costs than the individual hospitals could negotiate on their own. When VHA's robotic process automation champion learned that the organization's business operations staff was spending a substantial amount of time searching the Internet for product specification data, he pushed to automate the information-search process and to link it to procurement. The savings came quickly: In a few short months, the automated process pulled more than 360,000 product descriptions from the Internet, freeing staff to work on other activities related to sales and revenue generation. By targeting a painful and visible task, the company not only received buy-in from employees but also stirred enthusiasm for increased service automation in general.

Involve IT early. The IT department can be an important contributor to the success of an automation program. In our research, we learned that in several instances, champions, including the champion at Telefónica UK's O2, attempted to introduce service automation without involving IT. There were two main reasons: (1) executives saw service automation as a business operations program in that it required process and subject-matter expertise, not IT programming skills; and (2) they worried that IT would add too much bureaucracy and slow the rate of adoption. What some executives overlooked was the valuable knowledge IT can bring. IT can vet service automation software to ensure that it is safe, develop access rules to prevent software robots from exposing sensitive data, and maintain software robots on a safe, fully backed-up infrastructure. What we saw suggested that the pluses from including IT early far outweighed the minuses. As an executive at one service provider noted: "The minute we engage with business owners, we insist on speaking with the IT function. When we talk to IT, we explain that we have a product that is designed to appease their requirements for security, scalability, auditability, and change management."

Recognize that many employees are wary about the impacts of automation. Across our case studies, we saw companies using service automation tools to do repetitive and boring work. In the organizations we studied, the automation affected parts of jobs more than entire jobs,[14] and the effects on employment involved increases in productivity and reductions in hiring or outsourcing rather than layoffs of full-time employees.[15] Often companies redeployed internal employees to other business activities; service automation allowed them to avoid expanding their head counts. In fact, managers at the companies we studied reported that their employees were largely positive about the changes. Rather than feeling threatened by automation, many appreciated having fewer repetitive tasks and more opportunities to assume customer-facing responsibilities.

Nevertheless, it's common for employees to be apprehensive about the potential impacts of service automation on their jobs, and it's naive for executives to think otherwise. Prior research has found that communicating the intended effect on jobs early in the process is critical. In an information vacuum, employees tend to overestimate the ill effects; in some cases, staff members have panicked and even sabotaged new initiatives.[16]

Therefore, it's important for companies to be as forthcoming as possible about the implications for employees. Xchanging, the business-process provider, took an open approach to internal communications, using internal newsletters and regular presentations and road shows. These made robotic process automation developments visible quickly to everyone on the company's insurance staff. Management tried to make sure that operations teams were engaged in supporting projects and that they understood what service automation meant in terms of opportunities six to 12 months down the line.

Building Mature Service Automation Capabilities

For many companies in our study, the goal was to build an enterprise-wide automation capability. They expected automation to become part of the fabric of their business, much as computers and the Internet have become ingrained in organizational processes. Accomplishing this requires having a centralized command center that serves as a shared organizational resource. It requires organizations to rethink talent development and develop mechanisms for constant learning.

Establish a command center. A centralized command center helps business units across the organization identify automation opportunities, prioritize projects, build the solutions, and monitor the software robots once they take over tasks. A command center also establishes standards and best practices and tracks the business performance of service automation.

Among the organizations we studied, the European utility had the most mature service automation capability and provides the best example of what a centralized command center can be. The center (which the company referred to as its "center of excellence") managed the high demand for automation that came from customer transformation programs and from operational teams across all of the company's business divisions. When business operations teams proposed processes for automation, the center assessed their suitability and, if the project seemed promising, developed the business case. Once the case was approved and funded by management, the center was responsible for developing the automation solutions, testing them, and controlling the software robots once they were working on real data.

One of the best reasons for a centralized command center is that it can efficiently reuse software robots to scale quickly and to reduce development costs. For example, by reusing robots that were trained to log on to a particular system or to prepare a high-volume email from a customer database, the utility was able to reduce its development times by 30% to 40%. As one robotic process automation software provider explained, "The more processes you automate, the more objects you build in your robotic library. The more reuse you get, the more economic it is to assemble and deliver the new processes."

Rethink the talent development and skills needed for an enterprise automation capability. As organizations build automation capabilities, they need to rethink the skill sets needed to perform business services; different service automation roles require different skill sets. In the robotic process automation implementations we studied, companies added new roles, such as developers to build automation solutions and robotic process automation controllers to schedule, run, and monitor the software robots. For example, the utility company set out to recruit robotic process automation developers among people on its own operations staff who had a strong understanding of the business, process experience, and, preferably, systems analysis backgrounds. According to the company's robotic process automation project manager, the most important requirement was an ability to extract logical structures from disparate business data to build algorithms. IT skills were also seen as critical. He noted, "We're not IT staff, but we have staff with IT skills."

In contrast, for robotic process automation controllers to staff its control room, the utility targeted people who were organized, methodical, and logical and who had a consistent approach to work. It also sought people with good communication skills who could interact effectively with business operations people when they spotted any issues or anomalies. At peak times at the utility, two human controllers orchestrate the work of 300 software robots that do the equivalent of what more than 600 people once did.

Beyond considering the skills of the command center staff, it's also important to understand the capabilities of the retained human workforce. If robots are performing all of the repetitive and structured tasks, the humans will need to have more creativity, problem-solving skills, judgment, and emotional intelligence to tackle the unpredictable requirements of unstructured tasks.

Recently, there have been plenty of predictions about the effects of automation on the nature of human work. Some pundits have predicted that automation will take over more and more functions, leaving very few tasks for humans other than lawn mowing and hairdressing.[17] However, our research has led us to anticipate a different future for the automation of knowledge work.[18] In the next five years, we expect that more and more work groups will be composed of both humans and software robots, each performing tasks for which they are best suited. The robots will very quickly extract, consolidate, and rearrange data for humans to assess and act upon. Humans will deal with new business requirements (which humans may later teach to the software robots), troubleshoot and solve unstructured problems, positively envision services for customers, and build relationships with customers. We are already seeing some of this today, but going forward, robots won't need as much preconfiguration or as much detailed instruction as tools evolve and as robotic process automation moves to the cloud.

Of course, the field of service automation is progressing rapidly. Many case study participants told us that the next horizon would be tackling unstructured data with cognitive automation tools. They want software robots to read unstructured text, such as text messages or emails, and to decipher what the data means. Software robots are very fast; they have the ability to process huge amounts of data and present an interpretation almost instantly, which could enable a big step forward for customer service. In practice, it would mean that an agent on the phone with a customer could ask a software robot to mine huge quantities of data to help customers solve problems in a few seconds. The present state of service automation puts us on the path toward this vision.

Notes

1. For a brief history of the evolution of work from craft to mechanization, see J. Trevor, "Work and the Robot Revolution," March 1, 2016, https://www.sbs.ox.ac.uk/school/news/work-and-robot-revolution.

2. Among the developers of automation software are companies such as Blue Prism, Celaton, UiPath, Redwood Software, and Automation Anywhere.

3. M. Ford, *Rise of the Robots: Technology and the Threat of a Jobless Future* (New York: Basic Books, 2015).

4. D. Thompson, "A World Without Work," *The Atlantic*, July/August 2015, 50–61.

5. A detailed analysis of types of work that can be automated and where this is leading is provided in T. H. Davenport and J. Kirby, "Just How Smart Are Smart Machines?" *MIT Sloan Management Review* 57, no. 3 (Spring 2016): 21–25.

6. Our interviews at Xchanging were done in 2015. Computer Sciences Corp. of Tyson, Virginia, bought Xchanging on May 5, 2016. See "CSC Completes Xchanging Acquisition," press release, May 5, 2016, http://www.xchanging.com/news/csc-completes-xchanging-acquisition.

7. We asked all the early adopters to indicate the one-year return on investment (ROI) for each automation project. The lowest ROI reported was 30%; the most common responses were in the range of 40% to 60%. Respondents from one health insurance company reported a triple-digit ROI. We do not know the detailed parameters that companies used to calculate ROIs, but costs typically considered employee training, employee time required to build and operate the software robots, and software licensing fees. Benefits included savings on personnel costs, but none of the companies seemed to calculate a dollar value for improvements in service quality, service speed, or compliance.

8. These figures refer only to the company's United Kingdom retail division, where automation was first implemented.

9. We studied early adopters, and no other service automation sourcing options existed. Since early 2015, several advisory companies have developed robotic process automation practices that offer companies more options.

10. M. C. Lacity and L. P. Willcocks, *Advanced Outsourcing Practice: Rethinking ITO, BPO, and Cloud Services* (London: Palgrave, 2012).

11. Although the benefits of cloud services are obvious (particularly for small and medium-sized companies), robotic process automation companies we contacted said that no companies were using cloud services as of fall 2015.

12. For studies that look at standardization, see R. McIvor, M. McCracken, and M. McHugh, "Creating Outsourced Shared Services Arrangements: Lessons From the Public Sector," *European Management Journal* 29, no. 6 (December 2011): 448–461; M. Sako, "Technology Strategy and Management: Outsourcing Versus Shared Services," *Communications of the ACM* 53, no. 7 (July 2010): 27–29; M. J. Bidwell, "Politics and Firm Boundaries: How Organizational Structure, Group Interests, and Resources Affect Outsourcing," *Organization Science* 23, no. 6 (November/December 2012): 1622–1642; and M. C. Lacity and J. Fox, "Creating Global Shared Services: Lessons From Reuters," *MIS Quarterly Executive* 7, no. 1 (March 2008): 17–32. For a study that summarizes processes suitable for outsourcing, see Lacity and Willcocks, "Advanced Outsourcing Practice"; and for a study that looks at processes suitable for shared services, see J. D. McKeen and H. A. Smith, "Creating IT Shared Services," *Communications of the Association for Information Systems* 29 (2011): 645–656.
13. "The Impact of Robotic Process Automation on BPO," panel discussion at the Automation Innovation Conference, New York, December 10, 2014.
14. Some analysis suggests that tasks within jobs, rather than whole jobs, will be automated. Thus, the focus should be not on whole jobs but on activities and processes and how they can be reconstructed as a result of automation. See M. Chui, J. Manyika, and M. Miremadi, "Four Fundamentals of Workplace Automation," *McKinsey Quarterly* (November 2015): 1–9.
15. In the case of Telefónica O2, for example, head count at the service provider in India was reduced but full-time jobs in the United Kingdom were maintained.
16. See M. C. Lacity and J. W. Rottman, *Offshore Outsourcing of IT Work* (Basingstoke, UK: Palgrave 2008), 20–22.
17. An alternative view on the abiding value of multiple human capabilities is provided by T. H. Davenport and J. Kirby, *Only Humans Need Apply: Winners and Losers in the Age of Smart Machines* (New York: Harper Business, 2016); see also G. Colvin, *Humans Are Underrated: What High Achievers Know That Brilliant Machines Never Will* (New York: Portfolio/Penguin, 2015).

18. A comprehensive review of automation and the future of work studies, together with detailed projections about impacts on work design, outsourcing, and future processes, appears in L. P. Willcocks and M. C. Lacity, "Lessons and the Future of Automation and Work," chap. 10 in *Service Automation: Robots and the Future of Work* (Stratford-upon-Avon, UK: Steve Brookes Publishing, 2016).

11

Organizing for New Technologies

Rahul Kapoor and Thomas Klueter

The emergence of new technologies, while holding great promise for society, often threatens the viability of established companies. There are plenty of well-known examples of this, such as Eastman Kodak Co. and Polaroid Corp. and the advent of digital photography. In many of these cases, the core challenge for established companies stems not from a lack of recognition of or investments in emerging technologies. Instead, it stems from the challenge of commercializing an emerging technology whose economic attractiveness with respect to the company's existing business model is not at all obvious in the near term.

Today, managers in many prominent sectors, including autos, financial services, energy, and health care, face the challenge of pursuing emerging technologies that carry a high degree of uncertainty with regard to their economic viability and their companies' competitive position. However, what is sometimes lacking is an understanding, guided by systematic empirical evidence, of what managers can do to overcome this challenge. Our research offers some insights.

To understand the challenge, one first needs to recognize the distinction between the new idea (invention) and its subsequent commercialization through a product or service (innovation). This is important because, within established companies, the decision-making processes and logic governing inventions differ significantly from those governing commercialization. Engineering and scientific personnel typically drive inventions within new technological domains, whereas business development and marketing managers drive the subsequent commercialization. The guiding logic for inventions tends to be around the search for superior solutions to existing problems or unmet customer needs, while the guiding logic for commercialization tends to be around improving the competitiveness and profitability of the business.

These logics may be mutually consistent in the case of some emerging technologies (such as energy-efficient vehicles and wireless telephony). However, in other cases, they create organizational tensions stemming from business model considerations related to how established companies create and capture value. For example, engineers at Xerox Corp. in the 1970s pioneered key inventions in information technology (such as the graphical user interface and Ethernet computer networking). But Xerox did not aggressively pursue commercializing these innovations itself because the company's top management thought they did not fit with the company's existing copier-based business model.

We studied such tensions and possible solutions to them through a two-year field study of the pharmaceutical industry. This industry has witnessed an important technology shift fueled by the emergence of biotechnology-based therapeutics. However, despite the enormous promise, there has been and continues to be substantial uncertainty about when scientific discoveries will emerge, whether those discoveries will achieve clinical success, and how commercialized drugs will create value for the different actors. We collected detailed data on investments in research and drug development for the top 50 leading global pharmaceutical firms from 1989 to 2008, and we interviewed more than 20 industry experts. Detailed findings from our study were published in the *Academy of Management Journal*.

We focused our examination on two new biotechnologies that emerged in the late 1980s and that gathered lots of attention: monoclonal antibodies and gene therapy. Both technologies represented a radical departure from traditional chemistry-based therapies, faced a high degree of scientific and commercial uncertainty, and required established companies to invest in new competences. Despite the challenges, many established companies initiated research in both technologies and generated patented inventions. But we observed that the extent to which established companies' research investments led to drug development activities differed significantly between monoclonal antibodies and gene therapy. While inventions were being readily translated into downstream drug development and commercialization in monoclonal antibodies, that was not the case for gene therapy.

This can be explained in part by the fact that monoclonal antibodies and gene therapy differ significantly in their fit with the existing pharmaceutical business model. Monoclonal antibodies, like traditional chemistry-based drugs, are standardized treatments targeted at mass markets and prescribed over the long term, resulting in recurring costs for patients and insurers. As a result, monoclonal antibodies represent a good fit with the traditional pharmaceutical business model.

In contrast, gene therapies are typically one-off or significantly less frequent personalized treatments for patients with genetic disorders, administered by specialized physicians. Consider an application of gene therapy to treat hemophilia A and B. Such treatment is predicated on a one-time personalized injection, which not only threatens the existing market for hemophilia treatment but also presents a lack of clarity regarding how such a new treatment would be priced and reimbursed. Hence, gene therapy represents a case of a technology that disrupts the existing business model. As we confirmed through our interviews, it is this disruptive nature that made it difficult for gene therapy inventions within established companies to garner resources toward subsequent development and commercialization.

We also explored how managers could overcome this challenge—and discovered that the answer lies in the organizational design through which companies pursue emerging technologies. We found that gene therapy research that was conducted in-house or via external research contracts, where the established pharmaceutical company made the development and commercialization decisions, was less likely to move toward commercialization because of the conflict with the company's existing business model. Companies were more likely to pursue subsequent gene therapy development and commercialization when the research was conducted via alliances with startups and universities, or within a separate research unit that was acquired. In such situations, the decision making with respect to drug development and commercialization was structurally separated from the parent organization and involved outsiders from startups whose mental models differed from those of executives within established companies.

The implication of our research for executives is clear: When evaluating emerging technologies, managers should assess not only the new functionality and associated competences that their companies may need to develop but also whether the emerging technology has a significantly different customer value proposition and profit equation. (Think, for example, of digital imaging and photographic film manufacturers, gene therapy and pharmaceutical companies, or self-driving cars and automakers). The greater the incompatibility between the emerging technology's business model and the company's existing business model, the greater will be the organizational challenge that the company will face in commercializing it.

Managers can then mitigate this challenge by creating an organizational structure where resource allocation and decision making around emerging technologies are decoupled from the established company and involve outsiders with different mental models. This can be achieved through strategic alliances and acquisitions of startups or research units. Alliances may offer greater flexibility, whereas acquisitions may provide greater control over the technology and intellectual property. Alternatively, pioneering companies could create new units such as Amazon's Lab126 and Google X through aggressive hiring. However, such a structure often comes with an added burden for a company's leaders, since they must manage the competing demands of the core business and the emerging technology initiatives.

Today many businesses are confronted with disruptive technologies such as 3-D printing, artificial intelligence, cloud computing, the Internet of things, personalized medicine, and renewable energy. An important consideration for managers is to move beyond the decisions of whether and when to invest to the question of how to invest in such emerging technologies. While executives may initiate preliminary explorations in these technologies, they may be constrained by the logic and decision-making processes underlying inventions and their subsequent development and commercialization. But with appropriate organizational designs, executives can help sustain their companies' success—even in the face of an ever-shifting technology landscape.

12

Mastering the Digital Innovation Challenge

Fredrik Svahn, Lars Mathiassen, Rikard Lindgren, and Gerald C. Kane

In 2010, a small group of managers at the automaker Volvo Car Corp. assembled to craft a vision for the future involving wirelessly connected cars. They recognized that the company needed to renew its innovation capability to compete more effectively in an increasingly digital environment. Doing so, of course, was easier said than done. One problem was that many managers didn't see a need to innovate digitally. Volvo Cars was a car manufacturer, after all, and not a digital business. Others saw the need to engage in digital innovation, but they couldn't get their head around how to do so. How could they convince their colleagues, when they didn't necessarily have a clear vision of what the innovation outcome would be and the process itself appeared to be ambiguous? At the same time, Volvo Cars' automotive business was strong, raising additional concerns about how to innovate digitally while maintaining core competencies.

Many incumbent businesses share similar questions to those Volvo Cars faced in 2010. A recent report by *MIT Sloan Management Review* and Deloitte found that nearly 90% of managers surveyed report that their industry is likely to be disrupted by digital technologies, yet less than half report that their company is doing enough to prepare for this disruption. Our four-year research project at Volvo Cars offers insights into a challenge that established companies must master—and competing concerns that they must balance—as they pursue digital innovations.

Working on a strategy for connected cars, the executive team of Volvo Cars outlined a vision that would allow certain digital aspects of the car to be updated after the car was manufactured and sold. New digital technology would enhance users' experience and enable new revenue streams. By rethinking traditional automotive product development cycles, car connectivity could increase the pace of change. It would also allow the company to engage with external innovation ecosystems and sync with developments in consumer electronics.

The executive team realized this vision would not be easy to implement, nor was it mainly about creating new technological infrastructures. This bold vision stood in stark contrast to Volvo Cars' existing innovation practices and business models. Indeed, pursuing the digital innovation necessary for connected cars required fundamentally rethinking the organization, while also keeping the core business functioning efficiently. To chart a new way forward while maintaining the integrity and viability of its core business, Volvo Cars had to balance four sets of interrelated competing concerns.

New and Established Innovation Capabilities

Volvo Cars' first concern was how to balance the need to develop new capabilities for digital innovation yet still preserve the established innovation practices surrounding its core business. The company had traditionally invested substantially in innovation within product silos and multiyear time frames by relying on hierarchical structures and modular product architectures. Digital technologies, however, required a new way of thinking that cut across these specializations and moved forward more quickly than the company had ever done.

To achieve digital innovation, company executives realized they had to cross-fertilize the company's existing innovation environments and break away from its conventional product development practices. Although this transformation was necessary to leverage the new opportunities afforded by digital technology, it would require shifts in the company's capabilities, routines, and structures in fundamental ways that would affect Volvo Cars' identity and culture.

To manage those competing concerns, the executive team mandated a new initiative known as the Connectivity Hub, a cross-functional team tasked with developing new innovation capabilities for connected cars. The Connectivity Hub director, Mikael Gustavsson, noted, "The main job was to establish a new network that didn't reflect the existing organization. The Connectivity Hub was an opportunity to bring different parts of the firm to the same table. We didn't have an integrated forum where we could discuss those things." The Connectivity Hub orchestrated a broad internal debate about digital innovation and spearheaded efforts to prepare the organization for it.

The Connectivity Hub was set up as a temporary initiative so that it would not be perceived as a threat to existing organizational procedures. Nonetheless, the initiative at first generated substantial pushback. Resistance to the initiative occurred most intensely among middle managers, who felt torn between long-term visions requiring new capabilities and short-term commitments involving existing practices. This resistance was not unfounded: New product development at Volvo Cars customarily required product details to be frozen years in advance so that they could be implemented in production. However, making decisions about car connectivity features three years ahead would be impossible. These features could not be designed that far in advance; instead, they had to be generated through ongoing developments involving automakers, external developers, end users, and regulatory authorities. The Connectivity Hub had to figure out how Volvo Cars could foster such continuously ongoing innovation processes without compromising its ability to produce cars.

Process and Product Focus

At Volvo Cars, process innovation was traditionally associated with production efficiency and incremental product improvements. But now the company faced a very different challenge in that its digital features were not necessarily defined up front. Yet Volvo Cars' executive team did not believe its connected car vision would come to fruition unless digital features could be integrated with the physical car environment. New innovation processes had to be developed while still benefiting from the company's current strengths in building cars.

To manage these competing concerns, Volvo Cars explored how to develop generic digital resources, rather than simply focus on addressing specific end-user problems. Such generic resources offered prefabricated digital building blocks that could be utilized, combined, and built upon to resolve new innovation problems in the future. To legitimize such efforts in an environment inherently focused on specific functions, Volvo Cars built a portfolio of different platforms, each with a limited scope and distinct focus. These platforms were gradually developed to cover a broader range of applications. This approach allowed the automaker to shift its current focus on product platforms for cost-efficient implementation of predefined products to digital platforms that enabled new, often unforeseen, digital services.

As an example, Volvo On Call was originally a telematics service with specific features for remote car unlocking and safety monitoring. Volvo Cars realized this technology could be developed to issue generic digital keys that would enable retailers to deliver groceries to a specific vehicle. This service was later expanded, and the digital key is now a centerpiece in a commercial platform called In-Car Delivery, connecting car owners, logistics organizations, and a whole range of retailers in different niches in several European countries.

External and Internal Collaboration

When Volvo Cars started to conceptualize its digital capabilities as generic functions, questions were soon raised about who would use the different platforms to develop new services. The company had long controlled the internal collaboration required to leverage the scale advantages that its investments in modular product designs afforded through specialization and effective division of labor.

Not surprisingly, it became clear that this approach would not be able to release the potential of digital technology to produce increased variation and novelty of digital services for connected cars. The availability of digital platforms made the automaker realize the importance of also engaging external stakeholders as co-creators of value for the connected car aftermarket. Volvo Cars therefore launched a new software environment, called Volvo Cloud, to host in-car services based on software in back-end servers. This successful initiative opened up possibilities for external collaboration with third-party app developers, such as Pandora Internet radio and Spotify's digital music services, to secure a steady flow of new digital services to Volvo Cars' customers.

Because software resided in the cloud rather than being inscribed into car parts, the automaker could manage innovation concerns by opening up opportunities for collaboration with external partners without disturbing existing internal innovation practices. Volvo Cloud allowed in-car functionality, such as web radio, to be enabled, updated, or replaced without touching the car configuration.

Flexibility and Control

Collaborating with new partners on product development in turn invited new ways of thinking about innovation governance. Early collaborations with application development companies revealed that these new partners would not engage on the same terms as traditional suppliers of car parts. The automaker therefore launched an app development group, staffed with people from the consumer electronics industry and embedded within Volvo Cars' internal research and development department. This group served as a bridge between internal and external environments and crafted a range of boundary-spanning resources that could help Volvo Cars interact with external application developers. Initially, this spurred a series of cocreation initiatives involving external partners such as the location-sharing app Glympse.

These options, however, lost momentum during transition from demonstration to commercialization as soon as Volvo Cars'

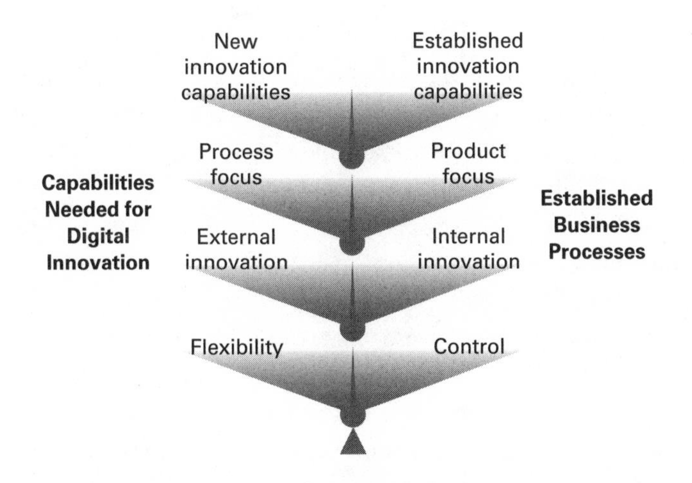

Balancing Competing Concerns in Digital Innovation
For established companies such as Volvo Cars, pursuing digital innovation means balancing competing concerns by fostering both new capabilities and existing core competencies. Specifically, Volvo had to strike a balance between: (1) new and existing innovation capabilities; (2) process and product focus; (3) external and internal innovation; and (4) flexibility and control in partnerships.

purchasing department got involved. The purchasing department staff instinctively applied traditional supplier contracts, based on monetary transactions, to regulate supplier implementation of Volvo Cars' requirements. However, in its collaborations with Pandora and Spotify, the automaker did not write elaborate requirement specifications or pay these partners—essentially making traditional contracts useless.

To effectively manage the relationships with these new types of external partners, Volvo Cars learned to balance the need to control the relationships with enough flexibility to stimulate value cocreation. In doing so, the automaker crafted a new contract that emphasized mutual liability and cost neutrality. This contract recognized partners' need for Volvo Cars' long-term commitment to support delivery of high-quality digital services and formalized, sustainable relationships that did not involve explicit monetary transactions.

Embracing Digital Innovation

Volvo Cars' journey toward digital innovation offers several lessons about how incumbent companies can compete more vibrantly in digitalized environments. First, it demonstrates that digital innovation is an organizational capability, not merely a new technological platform or an innovation incubator. Developing digital innovation capability requires fundamentally rethinking how the business is organized, how it makes decisions, with whom it partners, and how those partnerships are managed. These concerns are systematically interrelated and mutually dependent, so companies may find that a failure to address any of these competing concerns may have a wide-ranging impact on the overall success of digital innovation initiatives.

Second, it is possible for established companies to develop digital innovation capabilities while maintaining their core businesses. In fact, it is essential to do so. Successful established companies possess knowledge and expertise that have served them well for years, and the way they have done business is largely institutionalized. Digitalization provides opportunities to infuse new types of features and services into existing products. Doing so certainly means that some established ways of doing business must change. For example, this was true for Volvo Cars' traditional innovation cycle time, collaboration patterns, and partnership governance practices. This recognition, however, doesn't mean the old way of doing things is necessarily wrong. In fact, Volvo Cars managed to retain critical aspects of its approach to car design and production that still proved useful. Incumbent companies must find ways to leverage their strengths to capitalize on new ways of doing business. Ultimately, managers need to carefully identify which practices need to change and which need to be preserved.

Lastly, developing digital innovation capabilities will not happen by accident. Volvo Cars' executive team developed a clear vision for the broad parameters of their efforts, even though they did not yet know much of the specifics. They communicated this vision to the organization and provided the necessary support and resources to begin and endure the journey. Not everyone was initially on board. Many executives saw it as a high-risk, low-reward undertaking. Yet the clear communication of the vision, and the implementation of changes when needed, allowed Volvo Cars to innovate in its organization and its products to continue competing in an increasingly digital business environment.

This article is based on research that appeared in F. Svahn, L. Mathiassen, and R. Lindgren, "Embracing Digital Innovation in Incumbent Firms: How Volvo Cars Managed Competing Concerns," *MIS Quarterly* 41, no. 1 (March 2017): 239–253.

IV

New Approaches to Social Media

13

Finding the Right Role for Social Media in Innovation

Deborah L. Roberts and Frank T. Piller

Social media success stories have become widely shared narratives, highlighting the impact social media can have on companies' fortunes. For example, Burberry Group PLC, the London-based luxury fashion brand, relies heavily on social media to reach customers and fans.[1] As far back as 2011, Burberry was spending more than 60% of its marketing budget on digital media.[2] Increasingly, companies are attempting to navigate the social media landscape and use social media as a business tool to enhance performance. This is reflected in reports of increased spending on social media initiatives and the establishment by some organizations of dedicated social media functions.[3] Despite this, there is a significant opportunity that isn't being tapped: using social media to support innovation and new product development.[4]

Consultants and academics alike have been touting social media as a resource for innovation and new product development—a vehicle for developing customer insights, accessing knowledge, cocreating ideas and concepts with users, and supporting new product launches. Yet our research suggests that, despite the promise, the expected positive results are frequently not realized in practice. To begin with, the use of social media by companies for new product development lags far behind social media use by the general public. Although some companies have been able to use social media to develop new insights that lead to successful new products, many others simply do not know how to utilize social media for innovation. What's more, some companies have seen their innovation performance negatively affected. For instance, some get distracted by the diversity of input from social media; their traditional filters for screening data, like representativeness or consumer demographics, no longer work. Others waste resources by not validating the source and reliability of information; they mistakenly consider the information from social media to be just as valid as information from traditional online databases.

Nevertheless, we believe that social media provides a game-changing opportunity for companies that learn how to exploit it. But taking advantage of the opportunity requires more than having a Facebook presence with a loyal base of friends who say they "like" you. In order to use social media for innovation, organizations need clear strategies and objectives.

Several studies have examined how social media can be used for wider business purposes, especially communication, and as a driver of internal interaction and knowledge management within companies. In these realms, social media seems to have become an established part of the corporate tool kit.[5] Our research indicates that acceptance of social media for innovation has been less widespread. Fewer than 50% of companies surveyed use social media during the new product development process. Moreover, the use of social media tools can create unexpected challenges for managers. A study of northern European companies found that, for some managers, social media contributed to "infoglut" and made it difficult for managers to know which voices to listen to. In other words, there is a danger of listening to the wrong audience.[6]

About the Research

Our research is based on two original studies that examined the use of social media tools and sources within the new product development process of larger samples of companies. In the first we studied how a sample of 209 northern European companies used social media for new product development. Our second study utilized the 2012 Global Comparative Performance Assessment Study data set of the Product Development and Management Association (PDMA), the largest professional organization in the new product development field. This study benchmarks the practices of 453 companies (198 from North America, 149 from Asia, and 106 from Europe). Every few years, PDMA surveys hundreds of innovation professionals about best practices and success factors for innovation management, and the latest survey also asked about the use of social media for innovation. This data showed positive effects for companies that utilized social media in all stages of the innovation process, but it found that companies should have dedicated structures and an innovation culture to capture the benefits. In addition, our thinking about how social media can enhance innovation has been advanced by research and consulting projects with companies in Europe and the United States, in which we help our partners understand which organizational processes, structures, and cultures enable them to interact deeply with their customers via social media; which website design factors enable these conversations; and how companies can tap into unknown and unconventional sources of user-generated content using social media analysis.

We recently studied the social media practices of large global companies as they relate to new product development,[7] using data from the Product Development Management Association's 2012 Comparative Performance Assessment Study.[8] The data showed that 82% of the surveyed companies said that they have used social media for new product development. However, only 14.7% of the respondents said they used social media in at least 50% of their projects. Despite the hyperbole surrounding popular social networks, the findings showed that most companies actually use tools such as user forums and blogs much more frequently than they use sites such as Twitter and Pinterest.

Broadly speaking, we found that for many companies, the results of using social media for new product development fell short of expectations. Although social media did tend to help companies generate customer insights, companies that jumped on the social media bandwagon and invested in social media initiatives without a clear strategy, the right skills, or knowledge frequently did not achieve the results they were looking for. Those that utilized social media sources exclusively to search for technical information saw no improvements in new product development performance; in fact, the effect on performance for these companies was negative (due to information overload and the complexity of processing such information). The companies that benefited the most from using social media for new product development were those that used social media in every stage of the development process; they built organizational processes and structures to support new product development activity.

Before embarking on social media initiatives for new product development, managers need to develop a strategy and be sure they have the right processes and people in place to be successful. They need to figure out if their goal is to understand the latest trends in their marketplace and obtain customer insights, to cocreate with customers to develop new ideas and concepts, or to support the launch of their new products and use social media to create awareness and positive word of mouth among users. To illustrate the various approaches, we use the analogy of summer camp, a setting where children can explore and learn from everything they do. Like parents who aspire to find the right camp environment for their children, companies must consider their social media needs and strategies with care and intelligence. Companies that lack a vision of what they want to achieve by incorporating social media into their new product development process won't be able to reap the potential rewards.

For the purpose of illustration, we describe three different "camps": Camp Explore, Camp Cocreate, and Camp Communicate. Each camp offers a distinctive approach to thinking about the different phases of the innovation process and delivers an important skill set required to leverage social media for innovation. To realize the potential of social media for new product development, product developers must engage in three interrelated activities: (1) they need to listen to and learn from user-generated content; (2) they need to engage and facilitate dialogue with customer innovators; and (3) they need to find an audience of early adopters to create excitement for new products and collect feedback for their improvement. The three activities are not sequential but overlapping. Although listening and learning are important in the early stages of new product development, companies can use the ideas generated during the early stages to cocreate with customers later on. In addition, rather than simply scanning a few social media sites or looking for insights in Twitter feeds, companies need to pursue an integrated social media strategy that pulls together a wide range of different skills, capabilities, tools, and infrastructures.

Generating Customer Insights

When most people think about social media, they think primarily about well-known platforms such as Facebook and Twitter. In practice, however, there are many different types of social media, and it is often the lesser-known platforms, such as special user forums or expert blogs, that provide especially valuable information for innovation. For example, for companies innovating in the agricultural products market, there are specialized channels such as the British Farming Forum and the Pig Forum that provide a steady stream of information on topics related to breeding, feeding, and selling livestock. Moreover, there are many different types of users, ranging from active contributors and lead users to passive users (widely known as "lurkers"). To tap the potential of social media, innovators have to determine which skills and competencies they need to be effective in different stages of the new product development process. In the early stages, strong skills in market research and data analytics are critical, whereas in the later stages, the most essential skills are the ability to communicate with different types of potential buyers and the ability to understand and manage the impact of both positive and negative word of mouth.

Typically, obtaining data for new product development has been time consuming and labor intensive. However, the data and business intelligence made possible through social media have the potential to transform this area, providing information about trends in the marketplace, intelligence about competitors' products, and feedback on early concepts rapidly and at very low cost. For instance, bloggers share their ideas and opinions about almost anything with anyone prepared to listen, providing a wealth of information. Facebook and Twitter have facilitated an explosion in self-reporting among users of novel products and services, providing innovators with a huge repository of data. Tools that detect sentiments on products and brands are widely available. The data is both structured and unstructured and may be composed of mixed-media formats that include text, images, and social network information, thereby increasing their diversity and richness. The global scale and speed of information gathering in real time is extremely valuable, speeding up the process and reducing the time to market, while also providing access to unobvious information beyond the traditional search scope of the company.

Consider Nivea, the German personal care brand owned by Beiersdorf AG.[9] The Hamburg-based company is a worldwide leader in research on skin and body care. Yet when Nivea broadened its new product development processes beyond traditional market research techniques such as concept tests and focus groups, it saw significant benefits. For example, by analyzing user-generated content about its deodorants on Twitter, Facebook, and user forums, the company's development team got unbiased views about problems users were having with the products. Traditionally, product development in antiperspirants had focused on the length of protection, skin irritations, and scent. But social media users weren't concerned about those features at all—only about stains on their clothes caused by the deodorants' residues. The comments led to the formulation of Nivea Invisible for Black & White, the most successful new product launch in the company's history.

Camp Explore

If a company's objective is to identify market trends and to generate customer insights, the right program may be Camp Explore, where activities are designed to extend the breadth and depth of how organizations search for innovations. Here, people will learn to read the signals from large, diverse, disconnected, and unstructured pools of data generated by users. In addition, they will learn to analyze and convert blog posts, tweets, and user-generated content into valuable insights for new products. Specifically, they will need to acquire skills in computational techniques to unveil trends and patterns within and between the various data sets. Activities will encompass data analytics, machine learning, sentiment and textual analysis, data screening, evaluation, and data privacy. This camp offers experience in learning what can be automated, what requires the human touch, and how to interpret and make judgments about the data. Managers will have opportunities to develop the skills of both a social and a data scientist, so they can assimilate, combine, and utilize data from many different sources. The goal is to sharpen their business acumen and teach them how to communicate the findings to those involved in innovation projects.

Camp Cocreate

Companies that know they want to actively engage and involve customers in their innovation process should consider attending Camp Cocreate. The activities here are geared toward developing collaborative skills and facilitating interaction with users to involve them in the development of new products or services. Managers will learn how to work with customers and to cocreate value with them in the new product development process. They will learn how to engage, identify, and select the right participants and develop the right incentives to encourage their participation. Creativity is both an input and an output of the cocreation process. Managers will also develop skills in relationship building and gain experience in the art of conversation and dialogue, a key aspect of collaboration. Managers will learn how to become better facilitators and community managers. They will learn to develop and select ideas and product concepts that are suited to both target and nontarget market customers.

Earlier research has shown that ideas for new product development are often inspired by outliers or people from nontarget markets, so obtaining rich input from unconventional users can be an important factor to enhance the creative process. In this environment, managers can learn to use social media sites to post their own ideas as well as to explore what others are doing and vote on their favorite concepts. Instead of relying on one designer or developer, there's the opportunity to get dozens or even thousands of motivated users to engage with a task. To attract diverse input, some companies are adding apps to their Facebook pages; others are setting up their own ideation communities, which they either host themselves or through intermediaries such as Boston-based C Space or Munich-based Hyve Group, which maintain and manage dedicated cocreation communities. Relationships between companies and online communities don't just happen—they must be monitored and managed on an ongoing basis. Traditionally, managers set their own deadlines and work at a pace set by the company rather than by external entities. At Camp Cocreate, however, the party doesn't stop.

One good reason to consider Camp Cocreate is that several studies have supported the positive effects of collaborating with diverse participants in determining new product performance. For instance, when companies cocreate with customers, their products have been found to be more innovative and better suited to the market, thus making them more attractive to customers and enhancing profits.[10] Recent research has also found that good ideas and designs can come from both expert industrial users and end consumers.[11] Social media presents new opportunities for collaboration and idea generation, not just with selected users but also with a much larger network of users. Users and the communities that form around social media platforms can be sources of inspiration for new product development and sources of creativity in their own right. Social media allows individuals and communities to share, cocreate, discuss, and modify company- and user-generated content.[12]

In our own research, we have explored how to cocreate innovation using social media. A project with Ford Motor Co., for example, was aimed at creating innovative vehicle interior solutions for senior drivers, using a cocreation app on Ford's Facebook page. Although most of the ideas were aligned with concepts from Ford's own research and development team, some of the insights came directly from the users. In interviews, seniors told us that they appreciated the convenience and "flow" of using Facebook for exchanging ideas with other users.

Dell Inc.'s experience with IdeaStorm offers another example of a company using cocreation for innovation. Its original site, established in 2007, was an online suggestion box in which Dell customers could suggest ways the company could improve products, features, or support; at the time, this was a bold move.[13] In 2012, Dell added more advanced technology to allow greater collaboration and interaction with users. In a "Storm Session," company representatives engage users in real-time dialogue about a specific issue. IdeaStorm has served Dell well, both in terms of customer engagement and as a source for innovative ideas: More than 500 submitted ideas have been implemented.

Customizing Your Social Strategy to the Platform

Despite the potential benefits of engaging with users of social media and the communities that form around them, a number of studies show that companies are largely failing to realize benefits for innovation, irrespective of their social media presence or activity. In a recent study of 450 small- to medium-size companies in North America and Europe, we found that most (78%) use social media for marketing activities, and almost 40% of companies reported that Facebook was the most important online platform for their innovation activities. However, we found that what was important to the innovation process was not use itself but the approach taken. Companies that seek to use their social media capabilities to inform their innovation efforts should keep the following recommendations in mind:

Emphasize the social. The steep growth in popularity of social media is driven by people's innate human need for connection, communication, community, and social validation. People come together in these communities to debate issues, make new friends, and interact with family, friends, and business associates across the globe. Our research found that companies that recognized the importance of the "social" and helped to create an environment conducive to socializing—in other words, one that helped people to create or enhance relationships—benefited through people's subsequent engagement with the company's online innovation activities.

Customize your approach to each platform. To obtain valuable new product development insights from social media platforms, companies need to take a unique approach to each platform, since people use different social media platforms for different purposes. For instance, Facebook is predominantly a platform to enhance interpersonal relationships, while LinkedIn is primarily a vehicle for professional networking. Because Facebook is a more personal platform, people using it are more likely to be willing to share knowledge and ideas about products, brands, their own future needs, and insights on the competition. Their inhibitions are lower, and this can lead to more self-disclosure.

Professional platforms like LinkedIn, on the other hand, should be approached with a distinctly different tactic. On LinkedIn, companies can be more direct with individuals and groups when inquiring about products and features. For example, companies may have specific technology and product groups where specific design needs can be solicited from groups of engineers. Such community groups of specialized expert users can share ideas about new products or services and new markets—ideas that can be readily generated and shared without participants having to give away too much proprietary or personal information.

—Tucker J. Marion, Deborah L. Roberts, Marina Candi, and Gloria Barczak

---

Camp Communicate

Customer expectations have risen dramatically over the last decade, and companies often face the challenge of launching new products into crowded markets. Simply being innovative isn't sufficient; new products also need to be introduced in compelling ways. As social media becomes an ever more integral part of people's work and social lives, people have come to expect communication about products and brands via social media channels. In the past, information about new products was broadcasted to target markets in a linear fashion via paid advertising on television and radio and in newspapers. Social media, by contrast, consists of direct interactions with friends, peers, remote contacts, and the company developing the new product. Attaining positive affirmations (such as "likes" on Facebook or Twitter) can attract attention, which can stimulate interest in the new product launch. This, in turn, can lead to early product acceptance and subsequent demand. Social network sites can provide innovative and interactive means of communicating with customers and trigger interest in a new product.

Nestlé SA, for example, made good use of social media early in 2015 with the launch of its new Kit Kat "Celebrate the Breakers' Break" campaign to promote its Kit Kat chocolate bar. In addition to utilizing Twitter and YouTube, Nestlé had promotional hashtags molded directly into the chocolate of the Kit Kat bars, thus adding novelty and visibility to an extremely competitive market sector.

If the company's goal is to generate awareness and publicize the launch of new products, learning how to design and develop innovative product launch campaigns using social media needs to be a core activity. Activities at Camp Communicate will help managers take on this final and often costly stage of new product development. Camp Communicate emphasizes marketing communication skills: how to tell a story that resonates with the target market and, specifically, how to do this via Twitter or other easy-to-consume formats for mobile users. It trains managers in how to identify and connect with opinion leaders and early adopters in a way that resonates with their lifestyle (which may be dramatically different from interacting with conventional trade journalists or public relations agents). At Camp Communicate, managers learn that communication is multidirectional—a steady flow of arguments, comments, and modifications. They also learn that messages can be hijacked and subverted,[14] and that they are not—and *can't* be—in full control of communications. One important skill is learning how to manage the risks associated with negative word of mouth and how to use positive word of mouth to the company's advantage.

A Dedicated Strategy

As we saw in the late 1990s with the dot-com boom, many companies make the mistake of following the herd. In the case of social media, they embrace whatever social media sites and strategies are in vogue without developing a coherent strategy for tying their social media activity to new product development. Having a Facebook page, creating a brand community, or having a social media page dedicated to a new product launch will not, on its own, improve a company's innovation performance. Although we didn't ask about it directly, many of the companies we surveyed didn't seem to recognize the differences and functionalities of different social media platforms and media sources. However, companies need to recognize that there isn't one social medium—but numerous different platforms and networks. For example, communicating with Facebook fans may be a great way to rally the support of opinion leaders and brand fans when launching a new product. But if a company is looking for latent insights from lead users, it might be better off tapping into a user forum in a related area, where participants are discussing relevant problems in detail.

As we have noted, social media use does not automatically lead to improved performance in new product development. To achieve that, companies must develop a dedicated strategy that links social media to product development and to their corporate objectives. Managers need to question what they are trying to achieve. Are they seeking insights to develop novel concept ideas? Are they searching for technical information to enhance the company's technical problem-solving capabilities? Or do they want to enhance creativity by reaching out to users and customers and cocreating new ideas and concepts with them?

Social media can provide input for answering these types of questions. Rather than just eavesdropping on existing user content, many companies will want to engage with users in greater depth. To do so, product developers need to learn how to engage users and how to maintain a continuous conversation with them. This requires understanding the different types of social media and how they can be used in different ways.

To the extent that the effectiveness of social media for new product development is influenced by so many different skills and competences tied to different functional areas, departments, and individuals, it's critical that top leadership play an active role by encouraging cooperation and idea sharing among the various players. In some organizations, there may be the need for a "social media innovation leader" whose job is to align the different strategies and tools and help define a coherent social media strategy for new product development. The job would not only be to manage relationships with users and contributors (vital as this is) but also to manage the relationships among the various colleagues in the company's different social media camps.

Notes

1. M. Phan, R. Thomas, and K. Heine, "Social Media and Luxury Brand Management: The Case of Burberry," *Journal of Global Fashion Marketing* 2, no. 4 (November 2011): 213–222.
2. C. Barrett and T. Bradshaw, "Burberry in Step With Digital Age," Aug. 31, 2011, https://www.ft.com/content/70689408-d3f2-11e0-b7eb-00144feab49a.
3. V. Kumar and R. Mirchandani, "Increasing the ROI of Social Media Marketing," *MIT Sloan Management Review* 54, no. 1 (fall 2012): 55–61; and M. Mount and M. Garcia Martinez, "Rejuvenating a Brand Through Social Media," *MIT Sloan Management Review* 55, no. 4 (summer 2014): 14–16.

4. Social media can enhance innovation in a company in two ways. First, it is a source of unconventional knowledge and information from current customers, noncustomers, external experts, and also internal colleagues. Using social media to tap into these sources increases the scale and scope of search—a core strategy to increase innovation performance. Second, social media is a way to communicate innovation internally and to facilitate change and organizational innovation within a company. While the latter is crucial for the company's long-term survival, we focus on the former in this article.

5. G. C. Kane, M. Alavi, G. Labianca, and S. P. Borgatti, "What's Different About Social Media Networks? A Framework and Research Agenda," *MIS Quarterly* 38, no. 1 (March 2014): 274–304; and G. C. Kane, D. Palmer, A. N. Phillips, D. Kiron, and N. Buckley, "Moving Beyond Marketing: Generating Social Business Value across the Enterprise," *MIT Sloan Management Review*/Deloitte 2014 Social Business Global Executive Study and Research Project, http://sloanreview.mit.edu/projects/moving-beyond-marketing/.

6. D. L. Roberts and M. Candi, "Leveraging Social Network Sites in New Product Development: Opportunity or Hype?" *Journal of Product Innovation Management* 31, no. S1 (December 2014): 105–117; also see Kane et al., "Moving Beyond Marketing." Companies that are more mature in using social media for marketing and communications are also more actively using social media for innovation. In contrast, 71% of those companies that consider themselves as being in the early stage of adopting social media are using it not at all or only very rarely for new product development. Data were calculated using the *MIT Sloan Management Review* tool kit for Kane et al., "Moving Beyond Marketing"; see "2014 Social Business Interactive Tool," 2014, http://sloanreview.mit.edu/projects/moving-beyond-marketing/.

7. D. L. Roberts, F. Piller, and D. Luettgens, "Mapping the Scale and Scope of Social Media Tools for New Product Development Practice," unpublished manuscript.

8. The PDMA Comparative Performance Assessment Study is a broad international benchmarking survey conducted by the Product Development Management Association. For more information on this survey, see S. K. Markham and H. Lee, "Product Development and Management

Association's 2012 Comparative Performance Assessment Study," *Journal of Product Innovation Management* 30, no. 3 (May 2013): 408–429.
9. V. Bilgram, M. Bartl, and S. Biel, "Getting Closer to the Consumer—How Nivea Co-Creates New Products," *Marketing Review St. Gallen* 28, no. 1 (February 2011): 34–40.
10. B. Cassiman and R. Veugelers, "In Search of Complementarity in Innovation Strategy: Internal R&D and External Knowledge Acquisition," *Management Science* 52, no. 1 (January 2006): 68–82.
11. A. K. Chatterji and K. R. Fabrizio, "Using Users: When Does External Knowledge Enhance Corporate Product Innovation?" *Strategic Management Journal* 35, no. 10 (October 2014): 1427–1445; and E. von Hippel, S. Ogawa, and J. P. J. de Jong, "The Age of the Consumer-Innovator," *MIT Sloan Management Review* 53, no. 1 (Fall 2011): 27–35.
12. R. V. Kozinets, P.-Y. Dolbec, and A. Earley, "Netnographic Analysis: Understanding Culture Through Social Media Data," in "The SAGE Handbook of Qualitative Data Analysis," ed. U. Flick (London: Sage Publications, 2014): 262–276; and F. T. Piller, A. Vossen, and C. Ihl, "From Social Media to Social Product Development: The Impact of Social Media on Co-Creation of Innovation," *Die Unternehmung* 66, no. 1 (December 2011): 7–27.
13. B. L. Bayus, "Crowdsourcing New Product Ideas Over Time: An Analysis of the Dell IdeaStorm Community," *Management Science* 59, no. 1 (January 2013): 226–244; and S. Israel, "Dell Modernizes Ideastorm," *Forbes*, March 27, 2012, https://www.forbes.com/sites/shelisrael/2012/03/27/dell-modernizes-ideastorm/.
14. C. Heller Baird and G. Parasnis, "From Social Media to Social Customer Relationship Management," *Strategy & Leadership* 39, no. 5 (2011): 30–37.

14

Improving Analytics Capabilities through Crowdsourcing

Joseph Byrum and Alpheus Bingham

How does a company operating outside the major technology talent centers gain access to the most innovative data scientists that money can buy? Assuming you can't recruit the right data analysts to join your team full time, how do you tap into contractors with the knowledge and creativity you need outside your technical core? In a nutshell, this was the predicament Syngenta AG faced in 2008.

Syngenta, an agrochemical and seed company based in Basel, Switzerland, was formed in 2000 by the merger of the agribusiness units of Novartis and AstraZeneca. Among its more than 28,000 employees are more than 5,000 highly trained experts in biology, genetics, and organic chemistry, many of whom hold doctorates in their field. As a company, Syngenta's mission is to develop innovative crop solutions that enable farmers to grow basic food staples such as soybeans, corn, and wheat to feed the world's growing population as efficiently as possible. That means pushing the envelope on genetics.

For centuries, plant breeding has been a labor-intensive process that depended largely on trial and error. Farmers tested different seeds and cultivation techniques in an effort to find plants with the best yields and most desirable characteristics. Luck played a decisive role, as breeders relied heavily on intuition and guesswork to decide which varieties to cross-pollinate. To find the most successful variety of corn, for example, a breeder might have pollinated hundreds or even thousands of plants by hand to see what happened.

Syngenta had been involved in a large-scale version of trial-and-error research and development (R&D), conducting field tests on hundreds of thousands of plants each year in more than 150 locations around the world. But given that the results of experiments are often shaped by quirks and idiosyncrasies, it was sometimes difficult to draw meaningful conclusions. Did one plant grow taller than other plants because of a genetic trait, or was it because it received more water and more sunlight? With traditional research methods, the only way to find out was to invest significant amounts of time and money conducting large numbers of additional tests, which becomes an expensive proposition. Indeed, it takes seven years, on average, to move a new plant variety from the early testing stage to a full commercial product. When you are spending hundreds of millions of dollars each year on R&D for seeds and crop protection (Syngenta's R&D budget in 2015 was about $1.4 billion), even small savings have a big payoff.

Our idea[1] was to use data analytics to study a wide range of plant and seed varieties so we could identify the most desirable plants early and make optimal use of resources (everything from capital to labor, land, and time). What if we could make smarter choices at every stage of the breeding process? In breeding, the process begins with selecting promising parent plants, cross-breeding them, evaluating the offspring, and commercializing the variety that demonstrates a proven ability to outperform existing products. Constant testing and retesting is central to plant breeding, but what if we could eliminate the cost of testing and retesting varieties that weren't good enough and select the likely winners earlier? Rather than investing time and resources on more and more testing, our aim was to make decisions about our plant portfolios using hard data and science.

Syngenta's product research and development site in Slater, Iowa, just outside Des Moines, is well off the beaten track for many of the people whose analytics skills we hoped to tap. We knew that we had limited capacity to compete with employers such as Google Inc. or the US National Security Agency for people trained in analytics. So we figured that our best bet was to be creative—to augment our in-house resources by partnering with consultants and academics in fields unrelated to biology and agriculture.

Learning to Use Crowdsourcing Platforms

Open innovation can help companies tackle complex business problems that they can't solve on their own. In some cases, the barrier is a lack of expertise; in others, it's cost. However, leveraging the potential of outside experts requires close cooperation from in-house employees, who need to feel that it's good for the business and doesn't threaten their jobs. Cooperation from staff is also essential for framing problems and evaluating options. At Syngenta, we turned to several online crowdsourcing platforms to find talent that could help us increase our R&D efficiency.

But before we looked outside, we looked inside. Every problem, challenge, or contest was posted internally so that staff had the first opportunity to offer solutions. Even when in-house talent lacked the particular skill set needed to address complex mathematical issues, their practical experience in plant breeding helped refine the questions we were asking. Rather than seeing the shift as threatening, employees saw that their input was being used to advance an ambitious project.

We were trying to make the most of crowdsourcing platforms and also learning how we could leverage advanced mathematics to develop better varieties of plants. Although crowdsourcing was not new, we wanted to learn how to apply it to our circumstances. We realized early on in reviewing the available crowdsourcing websites that there were different types of platforms for different purposes. We tried several. Some of them were static: When you posted a problem online, individuals stepped forward with a solution. This worked well in identifying, for example, statistical approaches to plant breeding issues that could be solved by pure mathematics. However, static platforms were not well suited to solving more complex problems that cut across multiple disciplines. For these, more curated platforms proved useful by gathering experts from several fields into teams—where people could address problems iteratively. For instance, a plant's ability to adapt to different locations is driven by biological nuances that don't lend themselves to solutions that an individual operating on his or her own could develop using only mathematics.

At Syngenta, we set out to use open innovation to harness the power of data analytics so we could identify genetic combinations that unlock desirable characteristics in soybean plants, such as the highest yield. There is no one perfect soybean plant; rather, there are different varieties that are particularly well suited for different climates and growing conditions. Given that a soybean has 46,000 genes that determine its potential, and the number of possible combinations is practically infinite, identifying the best plants was a huge challenge. To find the best-performing soybean varieties, you needed to put them to the test, comparing how one seed performed against others grown in different conditions across multiple locations around the world.

Our vision was to create a suite of software tools that would replace intuition in plant breeding with data-backed science.[2] For the initial tool, we wanted a data monitoring system that enabled breeders to glance at data from a given field and know immediately what had happened. We established a contest on the platform of Waltham, Massachusetts-based InnoCentive, which hosts a diverse community of users, including mathematicians, physicists, and computer scientists, who are eager to put their problem-solving skills to the test. The contest was open to all of the platform's 375,000 users. We wanted contestants to create a tool to represent field test results visually, taking the raw data from field trials and highlighting the anomalies for further investigation.

The tool we envisioned would conduct what's known as a "residual analysis"—the calculated difference between the observed value of a genetic trait and the predicted value of that trait based on a statistical model across many locations. Since we were looking for a methodology, we intentionally targeted the broadest possible audience. We wanted to test and obtain as many creative options as possible, which meant sending the challenge to the widest possible network. We were pleased by the amount of attention our contest generated within the InnoCentive community. Over the course of about three months, more than 200 problem solvers downloaded the detailed problem description and data so that they could begin developing new ideas for tackling the problem. In order to participate, individuals agreed to sign an online nondisclosure agreement and to follow the contest's ground rules. In the first round, we were essentially asking them to submit a white paper outlining their approach to solving the problem. Our in-house staff members reviewed each online submission, and the reviewers found two different entries that solved the data quality problem. We picked the approach that we believed could be converted into a practical tool most easily. Under the agreed-upon rules, this was the participant who was paid.

From our perspective, we were building impressive analytics capabilities for our organization at a bargain—running several challenges cost us less than it would have to hire a mathematics professor to work in-house, for example. Each platform had template agreements that effectively lowered the cost of solving problems; we only paid for correct answers, usable methodologies, and the production of viable tools. We stated up front exactly what we were looking for, and participants understood that they would only be paid for giving us what we wanted.

With the first challenge's winning methodology in hand, we posted a second challenge on the same platform that shifted the focus toward improving the tool. We wanted to automate as many steps as possible so that data entry would be fast, easy, and intuitive. We also wanted the output to be as easy to understand as possible for plant breeders using the tool in the field. This time, contest participants came up with a system that visually represented the yield data from various field trials based on results; areas of low yields were shown in one color and areas of high yields in another. Based on this technique, scientists would be able to identify the problem areas more easily than they could from scanning columns of numbers.

The following example highlights how the data visualization tool worked. In one field, we saw that the results didn't align well with the expected growth patterns of the different seed varieties we were testing. Instead of the performance being determined by the variety, results appeared to be based on positioning in the field. Some rows underperformed on one side but performed as expected on the other side. We wondered what was going on. Thanks to the tool, our testing field's manager knew where to look. He discovered the reason for the anomaly: Due to a factory defect, the combine that had been used to measure performance in the field in question was not properly calibrated, so measurements were skewed on one side, but not the other. Thanks to the power of this analytics tool, the equipment manufacturer was able to correct a factory defect that, had it not been resolved, would have skewed testing results for anyone using this machine for years, wasting time and money.

Defining Problems for Problem Solvers

Encouraged by our initial efforts to use open innovation and analytics to improve our plant breeding efforts, we set out to build on our early success by defining a third challenge for problem solvers: identifying a mathematical approach to designing the most efficient experiments for evaluating plant yields. In plant breeding, scientists managing yield trials have three key decisions: how many varieties to test, how many locations to test them in, and how many times to repeat the test. Each of the three decisions recurs in three distinct test stages. Because there are so many possible combinations of these design variables (for our products, there are more than a trillion), breeders often sidestep complexity by supersizing their trials—that is, performing more and bigger tests on the assumption that better results will naturally follow. However, we knew from prior experience that "more" didn't always yield "better" results. In many cases, it just meant spending more time and money.

We knew that the nature of this problem was many times more complicated than the data quality issue. Data quality is something that could be treated for the most part as purely a mathematics problem. The problem solvers didn't need a deep understanding of biology to compete. The yield trial design challenge, on the other hand, could only be completed by those with both advanced quantitative skills and knowledge of biology. Thus, Syngenta turned to a different open-innovation platform that focused on teamwork.

Naturally, the team approach narrows the number of participants in the challenge. Only a handful of people had the subject-matter backgrounds to attempt a solution. The initial responses we received regarding the mathematical solution for managing trials were not what we had envisioned. We realized that the problem we wanted to address was more complicated than we originally thought. Despite our best efforts to frame the challenge clearly and unambiguously, participants inevitably interpreted it from their own perspectives, solving questions that we hadn't asked and focusing on elements of the problem description that, from our perspective, were less important.

It became a source of frustration on both sides: The problem solvers struggled to understand what we were looking for, and we had difficulty framing the question in a way that engaged people.

In our experience, one of the biggest challenges of open innovation is learning how to define problems you want to solve in ways that engage potential problem solvers. As a result, we have learned that rather than presenting problems broadly, it's often better to divide them into smaller chunks.

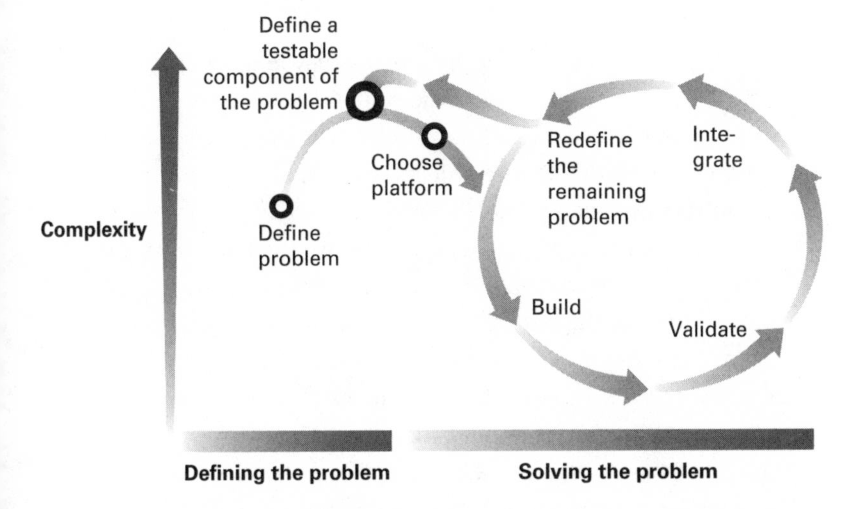

Solving Problems through Open Innovation

Rather than attempting to solve broad problems through open innovation, it's often better to divide the problems into smaller components. Potential solutions need to be built and validated. Valid models are then integrated into the ultimate solution, and the issue is redefined to explore another component. The process can be repeated until the entire desired solution is achieved.

In the yield trial design challenge, we realized that an approach that tackled the big picture (which would have required simulating over a trillion possible combinations for each yield trial) could take weeks and put too big a strain on computing resources. So we decided to redefine the problem, asking problem solvers to come up with a solution that would demand fewer resources. As much as we could, we removed biology from the question to focus on the mathematical component.

Validation is the critical step. In agriculture, we attempt to sort out whether the effects being modeled are the result of plant genetics or something in the environment; it's very easy to be led astray by false correlation. Validation is the only way we can know whether one part of a problem has been solved and that we are now ready to proceed to the next step.

One problem-solving team came up with a two-stage statistical approach that seemed promising. We tested it against the real-world, historical results we had. The mathematics held up when it was validated against the results, so we had a winner.

It took more than two years of testing and development to build the yield trial design optimizer. Along the way, we went through four major iterations and four minor revisions of the optimizer challenge on the open innovation platform until we arrived at the first version of our tool, fully validated and ready for use.

At this point, we knew that we were onto something big. We were raising our game in ways that wouldn't have been possible relying solely on in-house talent. The open-innovation platform gave us access to individuals able to develop the sorts of solutions that we never would have imagined possible. We have often found that the best answers come from unlikely places. For example, we worked with mathematicians and statisticians, as well as businessmen and engineers. Some of them were based many time zones away from Iowa—in Europe and even Australia.

Building New Capabilities for the Organization

As we became more familiar with how to manage input from outside experts, we found ways to engage outside problem solvers on a variety of issues. We have harnessed outside talent to come up with a tool that manages the genetic component of the breeding process—figuring out which soybean varieties to cross with one another and which breeding technique will most likely lead to success. We also used outside talent to create a tool that simulates the outcomes of all the different logistical choices our breeders might make—where to grow, which particular soybean traits are most important, and so on.

Over the past eight years, we have used open-innovation platforms to develop more than a dozen tools in our data analytics suite, which have cumulatively revolutionized the way we breed plants. By replacing guesswork with science, we are able to grow more with less, and that's exactly what needs to happen as the global population continues to increase.

We have found that there are several advantages to open innovation versus expanding our analytics capability in-house. First, since we look for people with mathematical and analytic expertise who can solve problems—as opposed to individuals who would also have the right fit with the organization—we haven't gotten bogged down in the traditional hiring process. When we found people who performed consistently, we signed contractual agreements to work with them directly. They became our "regulars."

So what's in it for the problems solvers? Amazingly, a lot of people participate in open-innovation contests because they enjoy the game. They might spend nights and weekends working to come up with a solution that ultimately doesn't work—and if they don't win, they don't get paid. One of our winning entries came from a man who owned a successful concrete manufacturing plant. He wasn't doing it for the money; his satisfaction came from solving a complex, real-world puzzle.

Of course, achieving results requires a great deal of effort. Managers who expect that working with online innovation platforms will be as easy as using online shopping websites will be frustrated and disappointed. Running a challenge requires more than answering a few questions on a form. For each contest, we formed a team of in-house experts to review submissions and communicate with participants. The team members shared a commitment to the ultimate goal of optimization because they had an idea of what success would look like. The tools we were building would make them much more effective at their jobs, but getting to that point wasn't easy.

Open innovation is a hands-on process. It involves frequent interaction with contest participants. When proposed solutions turned out to be more complicated than expected, we regrouped. Our in-house team would set up meetings with the problem solvers, either on the phone or in person, in an effort to better understand why something that seemed to our biologists to have a simple solution was more complicated from the perspective of a data scientist. We found in every instance that challenge participants who invested the effort to provide a submission were eager to work with us to redefine the problems we wanted to solve.

As a company, Syngenta has made a commitment to addressing the issue of global food security: A rapidly growing world population needs to eat, and that requires growing more food. Given the importance of this commitment, management recognized the company needed to upgrade its capabilities. Among other things, this meant learning to leverage data analytics in ways that had never before been attempted in our industry.

Before developing our new suite of tools, the average rate of improvement of our portfolio was 0.8 bushels per acre each year. Eight years later, we're looking at more than three times that—an annual improvement of 2.5 bushels per acre. Based on our analysis of our soybean crop portfolio, we estimate that we would have had to spend another $278 million on traditional field testing to achieve the equivalent level of genetic gains that we are realizing with the new tools. We are in the process of expanding what we have learned with soybeans to our entire range of crops, including sweet corn, field corn, sunflowers, and watermelon.

In April 2015, an independent panel of academic and business experts in operations research validated our analytics efforts and their applicability beyond agriculture, awarding Syngenta the prestigious 2015 Franz Edelman Award for Achievement in Operations Research and the Management Sciences.[3] What turned the judges in our favor? In the end, it was the combination of our ability to make better breeding decisions and the increased accuracy of measurements through all phases of the seed development and breeding process that paid off. The increase in yields was measurable.

In every industry, there's room for innovation—even if it means searching outside the company for new capabilities. Throughout the process, we have gone out of our way to explain open innovation to employees and also to show that our goal wasn't to undermine internal jobs. Employee input is essential in framing the challenges and evaluating the options. Our goal has always been to focus on the outcomes and to access the very best talent, wherever it happens to be.

Notes

1. Although Alpheus Bingham, one of the authors of this article, does not work for Syngenta, he served as an early stage advisor to Syngenta's project, so the authors opted to use the first person plural when describing Syngenta's experience.

2. More details about Syngenta's use of analytics can be found in J. Byrum, C. Davis, G. Doonan, T. Doubler, D. Foster, B. Luzzi, R. Mowers, C. Zinselmeier, J. Kloeber, D. Culhane, and S. Mack, "Advanced Analytics for Agricultural Product Development," *Interfaces* 46, no. 1 (January/February 2016): 5–17.

3. For more information about the Edelman Award, see http://www.informs.org/Recognize-Excellence/Franz-Edelman-Award.

15

Beyond Viral: Generating Sustainable Value from Social Media

Manuel Cebrian, Iyad Rahwan, and
Alex "Sandy" Pentland

We have witnessed social media playing a major role in mobilizing events of historic proportions, such as the Arab Spring protests in the Middle East and the Occupy Wall Street movement in the United States. Digital social media platforms, particularly Facebook and Twitter, are often cited as the facilitators of these mobilizations.

But most big social media-generated events seem to burst upon the scene, capture our attention for a few days, and then fade into oblivion with nothing substantial accomplished. No one—be they a charismatic leader or a raucous crowd—seems able to move people into action for extended periods of time using social media. This is especially ironic at a time when the online, crowdsourced society has reached maturity and is now widespread. Given all we have learned about social mobilization, why isn't it a more reliable channel for constructive social and long-term business change?

We argue that the rise of both social media and what author Moisés Naím has termed "the end of power" is anything but a coincidence. In fact, we view the confluence of these factors as a techno-social paradox of the 21st century.

We have studied why social media has provided the fuel for unpredictable, temporary mobilization, rather than steady, thoughtful, and sustainable change. In business, this may play out when a new product, company, or service—from phones to startups to games—grabs people's attention for a single announcement and then flames out.

How can businesses and others reverse this trend and reap more enduring benefits from social media? For starters, it will take a fundamental change in focus.

The Need for Incentives

We find that there is insufficient attention on the underlying incentive structures—the hidden network of interpersonal motivations and leadership—that provide the engine for collective decision making and actions.

A number of large-scale social mobilization experiments bear out the importance of incentive structures. Consider our own experience with the scientific scavenger hunt, the 2009 Defense Advanced Research Projects Agency (DARPA) Network Challenge. Our MIT Red Balloon Challenge Team competed with 57 other teams across the country to locate 10 weather balloons tethered at random locations all over the continental United States. None of the other teams managed to create a viral campaign that reached large populations and created awareness, while our team used an incentive scheme that motivated people to recruit their friends. As a result, we recruited over 2 million people in less than two days to help with the search, and we won the contest.

The difference in strategy was not just our emphasis on viral communications, but the way that incentives were matched with the motivations of the participants. Even for the simple task of finding balloons, successful teams tapped into people's motivation for personal profit, charity, reciprocity, or entertainment.

Our research shows that incentive networks are an important middle layer between ideologies and culture on the one hand, and the simple digital fingerprints left by social movements in online digital platforms such as Twitter and Facebook. They are part of what is fueling new areas of business such as the cocreation of products and brands through competitions and crowdsourcing.

Ideologies and culture shape what individuals want to achieve as they go about their daily lives, how they relate to each other's well-being, and how they help each other achieve those goals. These behaviors can be mapped into a network of incentives where each individual payoff depends on the payoffs of others. By contrast, the inability to sustain and transfer bursts of social mobilization into lasting social change or business results is rooted in the superficial design of today's digital social media—that is, it is designed primarily to maximize information propagation and virality through optimization of clicks and shares. However, this emphasis is detrimental to engagement and consensus-building. Understanding the dichotomy is an important lesson for those involved in online marketing.

Effective social media is the result of both information diffusion and recruitment incentives, yet most social media has focused on diffusion. From a business perspective, it means that social media is still extraordinarily ineffective at getting people to take action—for instance, by clicking through ads to make a purchase. As an industry, social media is still stuck on how to make people "like," not on establishing loyalty and stickiness.

One reason behind the emphasis on information virality is a phenomenon we call network measurability bias, which refers to the tendency to focus on processes that are easily observable within digital social networks, such as retweets. It neglects key latent processes such as the ideological, cultural, and economic incentives of actors.

But that's the wrong focus. If we shift our efforts toward mapping incentives, we may better determine the suitability of content for action—and create more lasting social and business change in the process.

Social media is an amazing tool that allows social scientists to measure information spread in real time, yet it is almost totally blind to other relevant factors, such as framing processes, reflection, consensus formation, or argumentation processes—which social scientists have found to be important in connecting content to sustained motivation.

Convincing someone of an idea is one thing. Recruiting them to incur costs of substantial time, effort, and risk toward supporting a cause or brand—or buying a product—requires much more commitment. We need new experimental paradigms and tools that spur individual, social, and cultural incentives in social mobilization processes.

Experimentation with these concepts may take time. Incentives are far less visible than message content, and a particular action often results from multiple incentives. When we do develop these models, however, they will help us develop a new generation of social media that can go beyond flash fads and viral memes toward consensual construction of sustained change.

16

When Employees Don't "Like" Their Employers on Social Media

Marie-Cécile Cervellon and Pamela Lirio

More than 2 billion people worldwide are users of social media, making it a logical platform for companies seeking to attract potential employees and engage consumers with their brands. In addition to sharing information on brand activities through official social media pages or accounts, organizations also are represented on social media through the private social media activity of employees. In their private lives, employees play multiple roles. They are free to share brand-related information, make comments endorsing the organization's brand, and display behaviors that are consistent (or at odds) with the brand values and promise. For companies, the social media behavior of employees represents both an opportunity and a risk.

When employees talk privately about their brands or the industries in which their companies operate, their comments often have more credibility with their network of contacts than when they speak about them in professional contexts.[1] Depending on the substance of their remarks, this can be a plus or a minus.[2] Many companies, including Patagonia Inc., an outdoor clothing and gear company based in Ventura, California; Société Générale, the Paris-based banking and financial services company; and Pernod Ricard, a Paris-based producer of wine and spirits, encourage their employees to become "brand ambassadors" to consumers and job candidates on social networks such as LinkedIn and share the company culture on Facebook and Twitter. Businesses such as L'Oréal, the cosmetics company, have even implemented programs to accompany employees, including top management, on their digital journeys and help them communicate creatively and efficiently on social media.[3]

However, our research shows that for many companies, the opportunity to use employees as brand ambassadors has been only partially tapped. Although employers expect their employees—especially younger ones—to follow the employer's brand on social media, share its brand links, recommend its products, and recommend the company to job candidates, we found that on the whole, employees displayed very low brand engagement on social media.

About the Research

Our insights on how employees engage in social media are based on two studies. The first looked at French, German, and Russian/Eastern European employees of a multinational company selling fast-moving consumer goods. The age of the survey respondents averaged 39.4 years, with a minimum age of 22 years and a maximum of 59 years. Overall, 77% of 353 respondents frequently used at least one social media platform; Google+ and Facebook were the most frequently used. Although employers may expect their employees (especially those who are members of Generation Y) to follow the employer brand on social media and share the employer brand links, recommend products, and recommend the company to job candidates, we found the surveyed employees actually showed very low brand engagement on social media. This held true for baby boomers as well as members of Generation X and Generation Y, even when their level of job satisfaction was high.

We discussed the results with the multinational company's CEO, marketing managers, digital managers, and online-community managers.

The results highlighted the importance of employee brand engagement to behavior consistent with brand values. Also, we found that the role employees perceive they have on social media was strongly related to their brand advocacy on social media. The French and German samples displayed cultural differences: French employees were more likely to separate their personal use of social media from their professional use. It is likely that encouraging employees to strengthen brand performance on social media will be more effective in countries, such as Germany, where mixing professional and personal information is more common.

We tested the validity of the first study with a second study of employees and managers in France working in private-sector companies with at least 50 employees. Respondents used at least one social media platform. Of the 252 respondents, 25.4% were managers, 60.3% nonmanagerial employees, and the rest top management (heads of departments and directors). Their average age was 42.8 years old. The most frequently used social media platforms were YouTube and Facebook, with a median usage between one hour and three hours per week.

The results of the second study supported the first study. The propensity of employees to engage in word of mouth was captured through three constructs: giving opinions, seeking opinions, and transmitting opinions. Social networks enabled dynamic diffusion of information, with a single person being a seeker, giver, and transmitter of information at once. The employees who naturally shared information online were the ones who were the most sensitive to the endorsement of their employer brands on social media. Also, in online discussions, the perceptions of contacts with similar characteristics ("profiles") influenced the credibility of the information shared within the group.

Results were discussed and interpreted with the input of digital managers, marketing managers, and human resource managers from five companies in three sectors. Through working sessions and an extensive search of the literature on employer branding and related constructs, we developed a set of recommendations to help organizations foster employee branding behaviors.

We acknowledge financial support from EDHEC Business School through the Bonduelle Chair and from the University of Montreal's Direction des affaires internationales (International Affairs Office).

At a European consumer goods company we studied, for example, less than half of the employees followed the employer's brand on social media. Managers at several companies we studied were surprised to learn that their employees were not following them on Facebook or other popular social media sites such as Twitter, Instagram, and LinkedIn. Yet when employees are not fans or supporters of the company's products, this can send an ambiguous message to employees' contacts and deprive the company of potential supporters.

So what can companies do? At a minimum, employers can remind employees that their behavior on social media can have negative consequences for the company. In addition, employees should be sensitized as to how their engagement on social media, such as "liking" their employers' posts or sharing the employers' achievements, can send positive messages to external stakeholders.

Unpacking Employee Branding

Corporate branding involves creating a unique image for the organization and its brand in the minds of key stakeholders. It reflects the organization's effort to deliver its promise consistently to employees (internal branding), potential employees (employer branding), and customers (external branding). There has been little scholarly work exploring the role of employees in the branding process outside the service sector, where the role of employees in delivering the brand promise to customers is paramount.[4] However, with increased personal use of social network sites, it is becoming apparent that every employee on social media has a relationship with key stakeholders, be they colleagues, current or future clients, suppliers, or potential job candidates. This reality drives the desire to have employees interacting in positive and constructive ways in relation to their employer's brand, both in professional and private spheres.

Employee branding is a process whereby employees internalize the company brand image and project that image to customers, job candidates, and other stakeholders.[5] It differs from *employer* branding (which aims to enhance the organization's image in order to attract and retain talented employees) and *internal* branding (which focuses on employee motivation to achieve organizational objectives and provide customer satisfaction).[6] What's more, employee branding goes beyond internal marketing in that it motivates employees to communicate the brand image to multiple stakeholders, as opposed to merely satisfying their own needs in an employee-customer interface.[7] We treat employee branding as the outcome of a process that begins with employees internalizing the brand and that leads them to endorse the brand externally with both customers and potential employees.

Whereas corporate branding delivers the organization's explicit promise to key stakeholders, employee branding conveys the promise when employees internalize it and endorse it either explicitly or implicitly through brand-consistent behaviors.

In the retail and hospitality sectors, research has shown that customer satisfaction is directly tied to employees' attitudes and behaviors.[8] In addition, employee branding is being examined increasingly in relation to human resources functions.[9] Although the research on employee branding for attracting job candidates is relatively new, a growing number of companies are encouraging their existing employees to use social media to attract new ones. At Pernod Ricard's wine division, for example, young hires post photos and information about their jobs to an Instagram account (@prfuturevintage).[10]

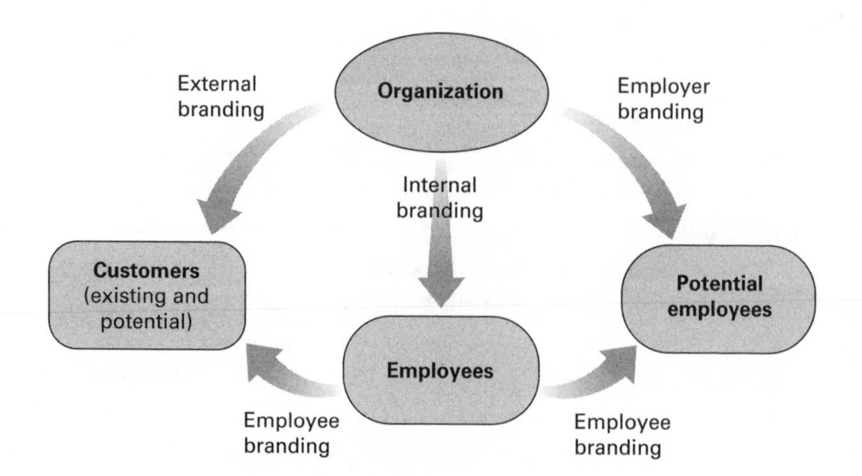

Understanding Employees' Role in Branding
Corporate branding delivers the explicit promise of the organization to key stakeholders. Employee branding conveys that promise to customers or potential hires when employees internalize the brand promise and endorse it explicitly or implicitly, through brand-consistent behaviors.

Employees explicitly endorse an employer brand on social media when they comment positively on the brand to their contacts, recommend the brand, share links, pass on information, or endorse the brand's values. Also, employees might refer to their brands indirectly by discussing relevant issues or through their behavior on social media. (For example, when a teacher discusses issues related to education on social media, his opinions might be seen as being endorsed by his employer even if the employer isn't specifically mentioned.)

For our research, we studied employee-branding behaviors on social media based on four dimensions: word of mouth, endorsement, sharing, and culture.

Assessing Employee Branding Behaviors on Social Media

Companies can map the extent of employee engagement on social media with their employer brand across four dimensions. Top management likely expects employees to be engaging in all 14 items across the four dimensions below. The reality, however, is probably different; in the companies we studied, the vast majority of employees engaged in fewer than seven of the 14 behaviors.

Dimension	Behaviors
Employee Word of Mouth	1. I speak (positively) about my employer brand on social media. 2. I praise the achievements of my employer brand on social media. 3. When I have a criticism about my employer, I refrain from sharing it online. 4. I do not post comments about my employer online that I might regret later.
Employee Endorsement	5. I am a fan/I follow my employer brand on social media. 6. I recommend my employer brand products and services to my contacts on social media. 7. I recommend my company to potential job candidates. 8. I respond (constructively) when my contacts criticize my employer brand products or services on social media.
Employee Sharing	9. I pass along information about my employer brand on social media. 10. I "like" content posted by my employer. 11. I share links to/from the employer brand on social media.
Employee Culture	12. I keep in mind that I could harm my employer brand when interacting on social media. 13. I behave on social media in a way that is consistent with my employer brand values and culture. 14. I communicate on topics related to my employer's business in a way my employer would approve of.

As a starting point, employers can conduct anonymous surveys about employee behavior on social media with regard to the company brand. This can sensitize employees to how their engagement on social media can benefit the employer brand. Surveys can be broadened to include evaluations of employee job satisfaction and employee voice, which we found to be two strong antecedents of willingness to participate in employee branding. However, we found that even when employee satisfaction is high and employee voice is valued inside the organization, most employees are not as engaged as we expected. This raises the question: What can companies do to encourage employees to become effective brand ambassadors?

Fostering Employee Branding on Social Media

We identified several factors that stand in the way of employee branding behaviors on social media. First, at the organizations we studied, there was often a lack of understanding among employees of the organization's social media strategy; many employees were even ignorant of their employers' social media activity. Second, employees were insufficiently aware of the importance of their role on social media. Most employees didn't know what was expected from them; one employee out of three could not say whether their company had a social media policy establishing do's and don'ts for them on social media. Third, there was discomfort around using social media in professional settings. This factor was more pronounced among senior employees (both in terms of age and position within the organization) and among employees who maintained a strict separation between their private and professional spheres. Based on these and other factors, we were able to develop a set of five best practices and recommendations for encouraging effective employee branding on social media.

Empower a stable of employee advocates. Certain groups are more conscious than others of the importance of endorsing their employer's brand on social media. While young and senior employees alike have embraced social media in their personal communications to some degree, those who were born in the era of the Internet—so-called "digital natives"—tend to be more active. Compared with Generation Xers or baby boomers, they typically maintain less separation between professional and personal information.[11] They are also more accustomed to voicing approval and disapproval on social media. Accordingly, young adults are more likely to become brand ambassadors for the company on social media. However, we found that an employee's age does not necessarily dictate the role he or she plays in promoting a brand on social media. As long as employees understand the role they can play on social media and how to engage with it, brand-building behaviors are similar across generations. Across our studies, we found that Generation X and baby boomer respondents used social media regularly, but they tended to be more comfortable using media passively (for example, reading posts) than they were sharing content or posting comments. Consequently, the first step toward building a collective digital culture may be to encourage digital natives to set the tone and help remove the psychological barriers senior employees feel toward social media engagement. This may involve "reverse mentoring," where younger employees help their colleagues increase their social media competence.[12] At L'Oréal, for example, digital natives work with senior employees to coach them on social media.[13]

Interestingly, the more overlap employees have in their profiles with others in their network, the more likely they are to display brand-building behaviors on social media.[14] Indeed, they are more likely to endorse the employer brand in their personal online interactions because they believe their contacts may have a similar interest in the employer brand. Moreover, they display brand-building behaviors naturally—not out of obligation. Many employee advocates on social media hold off on endorsing their employer brand until some of their contacts make the first move. A powerful example of this can be found in the response by employees and customers to the firing in 2014 of the popular CEO of Market Basket Inc., a Tewksbury, Massachusetts-based supermarket chain. By banding together online (#SaveMarketBasket, #ArtieT), employees and customers were instrumental in reinstating the CEO and preserving the company's distinctive organizational culture.[15] The Market Basket case highlights the potential effect of rallying key constituencies to endorse the employer brand.

Employees can also play an influential role in shaping the company's social media strategy. Initiatives on social media that unite employees in relation to their employers, such as employer groups championed by community managers, provide a forum for active advocates. Internal social network platforms, such as Yammer and Workplace by Facebook, can also be useful. Generally, companies should ask employee advocates for their feedback and include them in discussions on how to improve the impact of social media activities.

Outline the boundaries of employees' social media presence. Employees display brand-building behaviors when they understand their role in the branding process.[16] When they feel partly responsible for the company's success, they are willing to invest in activities to enhance the customer experience.[17] Similarly, when employees perceive they can play a role in the success of the brand online, they are willing to exhibit brand-building behaviors through their digital networks and on social media sites. Curiously, in our surveys, many employees said that they didn't think it necessary to "like" their employer brand or share posts with their contacts.

Once employees understand they have a role to play, they need to learn social media etiquette for when and how to mention the employer. Often, many employees aren't aware whether their company has a social media policy. Its guidelines should be outlined and communicated company-wide, with clear descriptions of what managers consider acceptable.[18] Bear in mind, too, that the employees' involvement in social media can become a liability. For example, in a company operating in the luxury sector, employees had to be reminded that the items they were proudly crafting and personalizing for prestigious clients had to be kept secret and could not be posted on Instagram. Also, spending hours on social media during the workday, including generating posts aimed at creating value for the employer brand, might send a negative message to external stakeholders. In a nutshell, employers should sensitize employees to the risks of blurring their professional and personal spheres on social media (specifically noting statements and behaviors to avoid and issues that employees shouldn't discuss).

In addition to pointing out risks, companies can benefit by training employees on the basics of social media. For example, many employees are not Twitter-literate. Many employees we observed or met were not endorsing their employer brand because they didn't really know how to do so. Some of their questions were fairly elementary ones on topics such as how to retweet and what hashtags are used for. Digital natives who were introduced to Facebook when they were teenagers are often amazed when their parents ask them to translate Twitter-speak into English. Rather than expecting older employees to use their own social media accounts to share information on the brand, they can be encouraged to simply comment on the company Facebook page or on sites such as Glassdoor, where employee feedback is posted anonymously. From instructing employees on the use of social media tools to educating them on social media metrics, different forms of training can be configured to support the digital brand-building activities of different categories of employees. Senior management should participate in the training sessions as well. L'Oréal, for example, trains all managers, including those at the very top, to become proficient in using digital tools.

Foster brand engagement. Organizations should foster employees' identification with their brand by encouraging employee brand commitment, defined as a psychological attachment and loyalty to the employer.[19] When employees demonstrate high levels of commitment to the employer, they have internalized the desired brand image. This combined sense of belonging and engagement becomes solidified through the employees' attachment to the job and to the brand, and by their developing a high level of trust in their ability to safely voice their opinions internally. Employee brand-building behaviors are rarely rewarded. However, they enhance organizational performance and help portray the workplace in a favorable light externally.[20]

Employee brand engagement encompasses both an emotional dimension, through emotional attachment, pride, and personal meaning,[21] and a rational dimension, through internalizing the values and understanding the heritage of the brand. Across our studies, we have found that employees who understand the brand promise and have an emotional attachment to their brand are more likely to invest in brand-building behaviors on social media. Training on brand building might be useful in making the brand platform available to everyone within the organization.

To foster brand engagement, management needs to establish clear expectations regarding employee behavior that are consistent with the "psychological contract." This contract needs to be grounded both in internal communication (via messages about what the organization feels is important) and in external communication (via awareness of corporate and brand-name communication efforts as conveyed through advertising, public relations, and social media). Here, too, internal networks such as Yammer or Workplace by Facebook are important; they help to create an employer brand community and promote a link between internal and external brand representations.

Having an employer brand community makes it easier for employees to endorse the brand on social media. Nurturing the brand community fosters a sense of belonging and serves to encourage employees to support the employer's strategic branding initiatives.[22] Within the brand communities of academic institutions we surveyed, the most popular posts were related to personal achievement (for example, how an employee did in a triathlon) or social gatherings (for example, an annual company summer barbecue or holiday party). Blurring the boundaries between personal and professional events helps employees spread information about the employer culture throughout social media.

Make content relevant and easy to share. Employees perceive their role in social media as important to the company as long as they see the brand as being active. Brand external communication not only affects employees' image of their company's brand, it also shapes their brand-building behaviors and encourages them to be active participants. Research shows that a perception of high quality in external communication such as advertisements can positively influence how employees identify with their employer brands; specifically, it motivates salespeople to devote more effort to the brand.[23] Thus, a brand that's perceived as being actively present on social media sends an implicit message to employees that social media is important to building the brand (which, in turn, encourages employees to communicate externally via these networks).

However, being present on social media, while important, is not sufficient. In order to turn employees into brand ambassadors, it's essential that the company have relevant content to share. Across our studies, many employees regarded their employers' social media content as not compelling enough to share. To address this problem, companies need to include their employees in brand content generation and invite them to be key participants in brand social media activities. At a minimum, content should be presented in a format that's easy to share and communicate in real time. For instance, one employee told us he had discovered an ad posted on YouTube by his employer only after it had already been posted for several weeks. Information needs to be updated frequently and shared with employees. In addition, companies can provide tools and assistance to help employees generate content. L'Oréal Canada, for example, encourages employees to develop creative content on social media using its "content factory," which maintains an online library of video tutorials, product pictures, product reviews, and testimonials to facilitate employee engagement with customers.

Reward employee voice. Research indicates that employees respond more positively to intrinsic psychological rewards such as public recognition than to extrinsic rewards such as bonuses, which can even have negative effects.[24] Moreover, employee branding on social media is effective only if the employee's voice is seen as authentic and sincere. The most effective rewards are straightforward but often overlooked. They include listening to employee feedback, paying attention to employee suggestions, and congratulating employees on their achievements.[25] The consequences of ignoring these potential rewards can be serious. For example, we spoke to a manager who did not feel she was adequately recognized for supporting her company brand on her personal blog. Her response was to tone down her testimonials and hold off from sharing company news via her blog and other social media platforms.

Many companies use extrinsic reward systems to encourage employees to participate in internal social networks. Some companies award employees points when they post comments or for the number of shares or "likes" they receive on their posts.[26] However, such systems carry a risk that employers will be seen as manipulating employee voice and intruding in employees' private lives. The power of employee brand building lies in giving employees freedom to express themselves within the boundaries outlined by the organization. For example, employers might find ways to link their social media advocacy to an incentive system for employee referrals. For instance, ShoreTel Inc., a telecommunications company based in Sunnyvale, California, tracks incoming candidates via links employees share through their personal social media accounts.[27] ShoreTel employees report valuing a privileged relationship on social media with top management, and this experience encouraged them to share or retweet information. However, such a system is only possible if top management leads and champions brand-building behaviors on social media.

If employers want employees to be constructive and engaged on behalf of their brands on social media, they need to respect the personal nature of how employees express themselves on social media. The company's interest in employee branding should not extend to policing employees' behaviors online or requiring access to their colleagues' social media profiles. If and when online-community managers encounter anonymous employee comments on sites such as Glassdoor, rather than be defensive, they can address the comments with transparency while emphasizing organizational safety. The foundation of employee branding is mutual trust and respect between employer and employees.

Implications for Companies

During our research working sessions, managers at several companies expressed concern that their employees were neither fans of their Facebook pages nor following their employer brand on Twitter, Instagram, or LinkedIn. In today's social media-focused environment, employees are often a valuable source of information for both customers and job candidates. At a time when organizations everywhere are encouraging customers and other constituencies to recommend their brands on social media, not being able to present the voice of your employees may communicate lackluster enthusiasm on the part of employees toward the company.

Past research indicates that organizations seeking to become leaders need to clearly state what is expected from employees and train them adequately on brand values and heritage.[28] Further, we recommend that companies find ways to integrate social media into internal branding strategies and training. In branding goods and services, the entire workforce needs to be trained to deliver the brand promise and engage actively with potential customers and job candidates on a day-to-day basis.

Notes

1. K. A. Keeling, P. J. McGoldrick, and H. Sadhu, "Staff Word-of-Mouth (SWOM) and Retail Employee Recruitment," *Journal of Retailing* 89, no. 1 (March 2013): 88–104.
2. S. J. Miles and W. G. Mangold, "Employee Voice: Untapped Resource or Social Media Time Bomb?" *Business Horizons* 57, no. 3 (May/June 2014): 401–411.
3. L. Lammers, "Patagonia's 'Tools for Grassroots Activists' Also Offers Lessons for Business, Marketing Leaders," February 29, 2016, http://www.sustainablebrands.com/news_and_views/marketing_comms/lesley_lammers/patagonias_tools_grassroots_activists_also_offers_less; V. Larroche, "L'Influence de la marque employeur sur l'E-réputation: L'Exemple de trois banques présentes sur le marché français" ("The Influence of Employer Branding on E-Reputation: An Example of Three Banks in the French Market"), in *Médias sociaux et relations publiques* (*Social Media and Public Relations*), ed. F. Charest, A. Lavigne, and C. Moumouni (Québec, Canada: Presses de l'Université du Québec, 2015), 65–82; N. Mortimer, "How Pernod Ricard Is Planning to Use Its Employees as Brand Ambassadors," September 4, 2016, http://www.thedrum.com/news/2016/09/04/how-pernod-ricard-planning-use-its-employees-brand-ambassadors; and J. Simpson, "How L'Oréal Uses Social Media to Increase Employee Engagement," October 22, 2015, https://econsultancy.com/blog/67091-how-l-oreal-uses-social-media-to-increase-employee-engagement.

4. N. J. Sirianni, M. J. Bitner, S. W. Brown, and N. Mandel, "Branded Service Encounters: Strategically Aligning Employee Behavior with the Brand Positioning," *Journal of Marketing* 77, no. 6 (November 2013): 108–123.
5. S. J. Miles and G. Mangold, "A Conceptualization of the Employee Branding Process," *Journal of Relationship Marketing* 3, no. 2–3 (2004): 65–87.
6. C. Foster, K. Punjaisri, and R. Cheng, "Exploring the Relationship Between Corporate, Internal, and Employer Branding," *Journal of Product & Brand Management* 19, no. 6 (2010): 401–409.
7. S. J. Miles and G. N. Muuka, "Employee Choice of Voice: A New Workplace Dynamic," *Journal of Applied Business Research* 27, no. 4 (July/August 2011): 91–104.
8. L. Xiong, C. King, and R. Piehler, "'That's Not My Job': Exploring the Employee Perspective in the Development of Brand Ambassadors," *International Journal of Hospitality Management* 35 (December 2013): 348–359.
9. W. F. Cascio, "Leveraging Employer Branding, Performance Management, and Human Resource Development to Enhance Employee Retention," *Human Resource Development International* 17, no. 2 (April 2014): 121–128.
10. Wine and Spirit Education Trust, "Graduate Wine Ambassador, Pernod Ricard Winemakers," n.d., http://www.wsetglobal.com.
11. P. Sánchez Abril, A. Levin, and A. Del Riego, "Blurred Boundaries: Social Media Privacy and the Twenty-First-Century Employee," *American Business Law Journal* 49, no. 1 (spring 2012): 63–124.
12. W. M. Murphy, "Reverse Mentoring at Work: Fostering Cross-Generational Learning and Developing Millennial Leaders," *Human Resource Management* 51, no. 4 (July/August 2012): 549–573.
13. L'Oréal human resources representatives, presentation to authors, December 2015.
14. S.-C. Chu and Y. Kim, "Determinants of Consumer Engagement in Electronic Word-of-Mouth (eWOM) in Social Networking Sites," *International Journal of Advertising* 30, no. 1 (2011): 47–75.
15. "Lessons From Market Basket: An MIT Sloan and Boston Review Roundtable," Oct. 8, 2014, http://bostonreview.net/us/lessons-from-market-basket-forum.

16. Xiong, King, and Piehler, "'That's Not My Job.'"
17. C. King and D. Grace, "Examining the Antecedents of Positive Employee Brand-Related Attitudes and Behaviours," *European Journal of Marketing* 46, no. 3–4 (2012): 469–488.
18. Sánchez Abril, Levin, and Del Riego, "Blurred Boundaries."
19. J. P. Meyer, D. J. Stanley, L. Herscovitch, and L. Topolnytsky, "Affective, Continuance, and Normative Commitment to the Organization: A Meta-Analysis of Antecedents, Correlates, and Consequences," *Journal of Vocational Behavior* 61, no. 1 (August 2002): 20–52.
20. C. R. Lages, "Employees' External Representation of Their Workplace: Key Antecedents," *Journal of Business Research* 65, no. 9 (September 2012): 1264–1272.
21. N. Bendapudi and V. Bendapudi, "Creating the Living Brand," Harvard Business Review 83, no. 5 (May 2005): 124–132, 154.
22. P. R. Devasagayam, C. L. Buff, T. W. Aurand, and K. M. Judson, "Building Brand Community Membership Within Organizations: A Viable Internal Branding Alternative?" *Journal of Product & Brand Management* 19, no. 3 (2010): 210–217.
23. D. E. Hughes, "This Ad's for You: The Indirect Effect of Advertising Perceptions on Salesperson Effort and Performance," *Journal of the Academy of Marketing Science* 41, no. 1 (January 2013): 1–18.
24. J. DiMicco, D. R. Millen, W. Geyer, C. Dugan, B. Brownholtz, and M. Muller, "Motivations for Social Networking at Work," in *Proceedings of the 2008 ACM Conference on Computer Supported Cooperative Work* (New York City: ACM, 2008), 711–720.
25. Miles and Muuka, "Employee Choice of Voice."
26. K. Ling, G. Beenen, P. Ludford, X. Wang, K. Chang, X. Li, D. Cosley, D. Frankowski, L. Terveen, A. M. Rashid, P. Resnick, and R. Kraut, "Using Social Psychology to Motivate Contributions to Online Communities," *Journal of Computer-Mediated Communication* 10, no. 4 (July 2005).
27. M. Feffer, "New Connections," *HR Magazine* 60, no. 3 (April 2015): 46–52.
28. W. G. Mangold and S. J. Miles, "The Employee Brand: Is Yours an All-Star?" *Business Horizons* 50, no. 5 (September/October 2007): 423–433.

V

Diving into the Void

17

Leading in an Unpredictable World

Pierre Nanterme, interviewed by Paul Michelman

Pierre Nanterme, chairman and CEO of the global professional services firm Accenture, has his hands full. He has set an ambitious agenda for his $31 billion enterprise to at once undertake its own digital transformation while also bringing a new set of digital capabilities to market in service of the firm's thousands of clients. As Nanterme describes it, Accenture is its own case study in digitization. In the spring of 2016, the Accenture chief sat down with *MIT Sloan Management Review* editor in chief Paul Michelman for a conversation about the challenges of digital transformation: for Nanterme's own organization, for leaders of the organizations he counsels, and for society. Edited and condensed highlights of that in-person conversation, as well as subsequent exchanges via email, are captured here.

MIT Sloan Management Review: You've written that leaders need to be both business savvy and technology savvy. On its surface, that's a statement that's very easy to agree with. But when we're thinking about senior leaders who may have come up at a time when technology did not loom nearly as large, what do we mean by "technology savvy?" What's the level of skills, the level of knowledge that a senior executive needs now?

Pierre Nanterme: The point is to have the right level of understanding. The future of the organization and the business model will involve combining business and technology opportunities. Digital technologies are becoming absolutely pervasive enablers of any new business model.

Now, the point is not for CEOs and other top leaders to be tech experts. The point is that you need to have a basic understanding of technology capabilities to figure out your new business models. This is certainly why many CEOs are spending time in Silicon Valley as an educational experience—to see use cases in action and to meet the next generation of leaders. And many of these tech leaders are already running the companies that are going to enable the CEO's next business model, which is going to be more facile, more global, cheaper, and more scalable.

And there's another important point here—and this is what's unique with the current wave of digital technologies—they can massively rationalize your operations. I do not remember many disruptions, if you will, which are enablers of new business as well as enablers of your own company's rationalization.

Let's stay on that theme. You've spoken about creating a "digital first" organization in two directions at Accenture: on the client-facing side and within the organization itself. And I think that challenge rings very familiar to executives; we're all struggling with how to respond to the market in a digital world, while also trying to future-proof our organizations. What have you learned from taking on both of these challenges simultaneously? Where are they mutually reinforcing, and where do they diverge?

They are mutually reinforcing. What I've tried to do leading Accenture is to consistently take the angle "digital first," and to be obsessed that we're going to bring digital into everything we do every day. Because indeed, that's going to enable the services we're going to bring to clients: what we're calling "the new." We specifically define "the new" as a set of five capabilities: interactive, mobility, analytics, cloud, and security. These are both the digital capabilities that enable our own evolution as a company and the capabilities we're going to bring to the market.

This is happening now. We have digitalized our governance: We're operating without any headquarters, without almost any in-person corporate meetings. Even me. I've been digitalized in the form of a hologram for large events. So, when I talk about digitization, I mean it literally. We've digitalized all our corporate processes: finance, human resources, training, communications, and so on. And we did so in a very deliberate way, based on our belief that the digitization wave is the major disruption of our time. It will reinvent what we are proposing to our clients and how we operate the business. What's true for us is true for our clients.

So are you finding that the firm is a lab for what you can then deliver to the market?

It's more than a lab. It's a case study. We are making Accenture a case for our clients. We are a giant. We operate in 120 countries. We have over 375,000 employees, so we have all the challenges of a large, global corporation. We are no different from the companies we serve.

When I'm talking with other CEOs, half of the discussion is peer to peer. You're leading a large corporation; I'm leading a large corporation. Tell me what you do—how you organize your global governance. I understand you have no headquarters; I understand your leadership team is distributed around the world. That's great. You're closer to the market. How are you enabling that?

And the other 50% is about my job as CEO, which is to propose our services. Being a digital services provider as well as one of the most digitalized large global companies creates massive synergies and huge credibility in my dialogue with other CEOs.

How have the dynamics of those conversations changed?
It's changed dramatically—and certainly with an acceleration these last six quarters. I think it's in correlation with the level of acceptance of the digital disruption by our clients. We had the 2008 to 2010 crisis, then the 2010 to 2013 recovery, where the focus was productivity, tactical cost reduction, and capital requirements.

And then, after 2013, digital is starting to kick, but it's initially kicking as a technology—say, putting infrastructure in the cloud—so it's for the chief information officer to manage. Or, yes, I understand digital is to have a website. So, I have a website, and I put my server in the cloud, and look at me, I'm digital.

That was the mindset in 2012, 2013, even into 2014. And then suddenly, let's say 2014 and certainly 2015, businesses started to figure it out. It was probably with the rise of Airbnb and Uber that suddenly, CEOs were realizing that what they conceived of as "being digital" was just the tip of the iceberg.

And only in the past 18 months, leaders are saying, "What we're talking about is a massive disruption. It's going to challenge fundamentally the existing business models. I could disappear because of the thing I didn't really get. Now I'm getting it." Suddenly, when you talk about digital, it's the CEOs you're talking to. Companies are appointing chief digital officers, and they're reporting directly to the CEO.

"All CEOs, including me, are operating in a world where it's almost impossible to predict things in the short term, and you need to be prepared to navigate."

So in spite of all of the hoopla around concepts – big data, the cloud, analytics – it took the emergence of new companies upending seemingly nondigital industries like transportation and travel to really get CEOs' attention.
Yes. I think for the last 10 years, digital has been a productivity game for most companies. And now that's not enough. It's not about being more productive in what you do. It's about fundamentally changing the way you produce, execute, and deliver the services and the experiences you provide, because technologies are enabling new entrants to take your business for a fraction of the cost. It's become a totally different dialogue.

In a similar vein, you wrote that success is no longer about changing strategies more often but about having the ability to execute multiple strategies concurrently. Can you expand on that statement? And what about digitization has wrought that?
It's about agility. Because all CEOs, including me, are operating in a world where it's almost impossible to predict things in the short term, and you need to be prepared to navigate. Of course, you need to have a direction that you believe is correct. We set the direction for Accenture to lead in providing digital-related services.

Now, we need to map out how we're going to get to the destination we've set for the business, even while we know the world will continue to shift during our travels. There will be new technologies that shape, even transform, the business landscape, so we will need to execute multiple strategies—or pathways—in order to make sure we get to where we want to go.

We'll use the metaphor of an ocean crossing. Not too long ago, I just needed one boat to make a voyage. And that boat had a predetermined course from which we would not need to veer.

Today, I need a fleet of speedboats, each following different paths across the sea. Some are fast, some are slow, and some are figuring out the nature of their boat. Where I'm going to land, I don't know exactly. It might be Miami or it might be New York, but I know I need to cross the ocean. I know generally where the future is, but I can't afford to be more absolute in my path than I am at this point in the voyage.

What's new—and what's making management so uncomfortable—is leading amidst this massive unpredictability. Even without digitization, you are dealing with volatile markets and macroeconomic factors. Together, it's incredibly destabilizing.

Leaders need to be prepared to reinvent along their journeys. You set your direction clearly and you understand that you will need to figure out almost every month what that direction requires. What are the capabilities you need to grow or to acquire? Where do you need to turn and to adjust?

That's the art of navigation: You know the destination, but you have no map.

How do you, as the CEO of a massive global enterprise that itself has to manage with this challenge, simultaneously oversee relationships with thousands of clients, each of whom is also dealing with this challenge? There's infinite complexity. How do you stay focused?

You need to understand the complexity of things and then oversimplify the execution.

So you put all of what we've discussed on the table, and you say, "Wow." OK, then we'd better start structuring. Let's begin by putting things in the right order. In order to reach our destination, we need philosophy, principle, architecture, investment, and so forth.

For instance, we are recreating the business architecture at Accenture. We now have five businesses—built around our new capabilities. We focus on five things. Not three. Not seven.

Of course, we should always take a look at new ideas, because one might turn out to be a big one, but first we are going to scale, especially when we have evidence that we have found a gold mine.

There are things you're making predictable even when the precise destination is still unknown. And this is what we are doing in a very deliberate way.

I'm assuming that you're giving similar advice to your peers out in the world.

It's exactly the same. We say to our clients, "Now you understand all the possibilities. It's time to make choices and to stick with it." Because you will have to align your investment, align your communication, align your training, and stand up for the direction you choose. It has to be an obsession: Stick and scale. And when I say "obsession," I mean it. Because you will have so many distractions, so many people asking, "Are you sure you're doing the right thing?" You need to be deliberate. "Yes, I stick, I scale, and I go for it."

You have said that "businesses could be more incentivized to assess the social impact of their digital investments." You believe that digitization and social impact have kind of a special relationship that is different from business activities in the predigital age.

Absolutely. The latest wave of the digital revolution will change people's lives. Life is not about productivity gains or privacy issues, two of the issues we've associated with technology advancements over the past 10 or 20 years. But with artificial intelligence [AI], cognitive computing, and machine learning, we are talking about things that can enhance humanity. And that's a different game.

But these technologies raise lots of difficult issues and concerns.

For example, there is a whole set of ethical issues in the way we are going to combine technology and the human mind and body—not just having them work together but how they physically interact. When we introduced the computer, it didn't change you physically. Now we're talking about a world where there's going to be less disease, where people could live much longer and in good health. And who knows how technology may enhance the brain itself.

Is it the end of days, or is it the beginning of new days? I believe it's the beginning of new days. I believe in progress. I believe that all of this is good.

But the concerns are real, and that's why we need to have a better understanding of the value digital technology could unleash for society—to have a truly balanced conversation.

We can reasonably begin to measure the value of a self-driving car or a new battery technology on business. But for society, what does it mean? It means that people unable to drive will be mobile again. If they are mobile, they can do things. And what they're going to do will have a value.

My parents reached a point when they couldn't drive. But if tomorrow they could drive, they could move. They could move from one place to another, and that place could be a movie or it could be a job. Either way, this is a profound change.

The mobility of people and products is the essence of trade. So when you have mobility back, especially when people are living longer, more value is being created. What is the economic value of living 20 more years, especially mobile years? You can contribute to the economy longer. You can work longer. You're going to consume longer. In our work with the World Economic Forum, we attempted to measure the value these new digital models could unleash. Our analysis put the figure at $100 trillion.

Yes, there are lots of issues, and there are times new technologies look more evil than good, but there is a disproportionate value for society.

But even as all of this new value is created, there's an inevitability that some workers will be displaced, right?
Yes, of course, but it's not universal.

One thing we do know is that technologies like AI are not going to affect all regions or sectors the same way. The efficiency gains and growth opportunities will vary by industry and by costs of labor. You need an economic equation that justifies replacing man with machine. I'm not sure that equation is going to be met in a lot of cases.

We have about 1 billion people in the United States and Europe, and more than 6 billion in the rest of the world. In the US and Europe, there will be disruption—but this is where you have the best education and training. This is where we—companies and governments—can best manage the trade-offs and provide the training for people to move into new added-value jobs. We can educate our people for the new economy. We have the resources.

But when you go to many other places in the world, perhaps you will never see any robots because there are not going to be any economic equations that make sense.

“Are you curious?
Are you keen to learn?
Are you prepared to
jump into the new?
Are you prepared
to lead in volatility
and ambiguity?
Those are the traits
of the new leaders.”

Accenture is very global. I'm traveling all around the world. And so I need to be careful to consider the concerns I'm hearing about in Boston in the right context. Go and visit Indonesia, and maybe you will come back and have a different view. Or go to Kenya or go to Mozambique, Angola, even some parts of India. And I think there is still a great deal of work that will be done by man.

We need to look at the world with a fresh perspective. And for the role of management, that means something important. We need to get past evaluating people on their numbers. I have a company full of people who can make their numbers. They've all been to good schools and they know how to get the job done; they all know the recipes.

What I care about now is: Are you curious? Are you keen to learn? Are you prepared to jump into the new? Are you prepared to lead in volatility and ambiguity? Those are the traits of the new leaders.

I find the unknown and the complex more fun than scary. If it's the other way around for your leadership, then you have a problem.

We are at the end of a management cycle that has covered the past 50 years, and we are beginning to write a new book for the next 50. There will be no long-term predictability. Leaders will need to learn to manage chaos and to do it in a highly disciplined way.

So if you ask me what do you see for your company in 10 years? I see nothing. It's too far out. I don't have a clue, and I don't care. But the next 18 months are crystal clear.

Contributors

Allan Alter is senior research fellow at the Accenture Institute for High Performance.

Stephen J. Andriole is the Thomas G. Labrecque Professor of Business Technology at Villanova University in Villanova, Pennsylvania, and is the author of the book *Ready Technology: Fast-Tracking New Business Technologies* (CRC Press, 2014).

Bart Baesens is a professor at KU Leuven in Leuven, Belgium, and a lecturer at the University of Southampton School of Management in Southampton, United Kingdom; he is also the author of the book *Analytics in a Big Data World: The Essential Guide to Data Science and Its Applications* (John Wiley & Sons, 2014).

Gloria Barczak is a professor of marketing at the D'Amore-McKim School of Business at Northeastern University in Boston, Massachusetts.

Cynthia M. Beath is a professor emerita of information systems at the McCombs School of Business at the University of Texas at Austin.

Alpheus Bingham is cofounder of InnoCentive Inc., a crowd-sourcing company based in Waltham, Massachusetts.

Didier Bonnet is an executive vice president and global practice leader at Capgemini Consulting and coauthor of *Leading Digital: Turning Technology into Business Transformation* (Harvard Business Review Press, 2014).

Chris Brady is a professor of management studies at the Centre for Sports Business at the University of Salford in Salford, United Kingdom.

Joseph Byrum is the global head of soybean research and development at Syngenta AG's US subsidiary in West Des Moines, Iowa.

Marina Candi is an associate professor at Reykjavik University's School of Business in Reykjavik, Iceland, as well as director of the Reykjavik University Centre for Research on Innovation and Entrepreneurship.

Manuel Cebrian is principal research scientist with the Data61 Unit at the Commonwealth Scientific and Industrial Research Organisation (CSIRO), Australia.

Marie-Cécile Cervellon is a professor of marketing at EDHEC Business School in Nice, France.

Simon Chadwick is a professor of sports enterprise and codirector of the Centre for Sports Business at the University of Salford.

Sophie De Winne is an associate professor at KU Leuven in Flanders, Belgium.

Mike Forde is a consultant specializing in performance and talent management for professional sports teams.

Gerald C. Kane is an associate professor of information systems at Boston College's Carroll School of Management and is *MIT Sloan Management Review*'s guest editor for its Digital Business Initiative.

Rahul Kapoor is an associate professor of management at the Wharton School of the University of Pennsylvania in Philadelphia, Pennsylvania.

David Kiron is the executive editor of *MIT Sloan Management Review*'s Big Ideas initiatives.

Thomas Klueter is an assistant professor of entrepreneurship at IESE Business School at the University of Navarra in Barcelona, Spain.

Mary C. Lacity is the Curators' Distinguished Professor at the University of Missouri–St. Louis College of Business Administration in St. Louis, Missouri.

Rikard Lindgren is a professor of informatics at the University of Gothenburg as well as research director and cofounder of the Swedish Center for Digital Innovation.

Pamela Lirio is an assistant professor of international human resource management at the University of Montreal's School of Industrial Relations in Montreal, Canada.

Tucker J. Marion is an associate professor in the entrepreneurship and innovation group at the D'Amore-McKim School of Business at Northeastern University in Boston, Massachusetts.

Lars Mathiassen is Georgia Research Alliance Eminent Scholar and a professor of computer information systems at the J. Mack Robinson College of Business at Georgia State University in Atlanta, Georgia.

Pete Maulik is a managing partner and chief growth officer at the innovation strategy firm Fahrenheit 212, a Capgemini Consulting company.

Paul Michelman is editor in chief of *MIT Sloan Management Review*.

Narendra Mulani is Accenture's chief analytics officer.

Pierre Nanterme is chairman and CEO of the global professional services firm Accenture.

Doug Palmer is a principal in the digital business and strategy practice of Deloitte Digital.

Alex "Sandy" Pentland directs the MIT Connection Science and Human Dynamics labs and previously helped create and direct the MIT Media Lab and the Media Lab Asia in India.

Anh Nguyen Phillips is a senior manager with Deloitte Services LP.

Frank T. Piller is a professor of innovation management at RWTH Aachen University in Aachen, Germany.

Iyad Rahwan is an associate professor of Media Arts and Sciences at the MIT Media Lab.

Deborah L. Roberts is an associate professor of marketing at Nottingham University Business School in Nottingham, United Kingdom.

Jeanne W. Ross is a principal research scientist at the MIT Center for Information Systems Research.

Ina M. Sebastian is a research associate at the MIT Center for Information Systems Research.

Luc Sels is a professor and dean of the faculty of economics and business at KU Leuven in Flanders, Belgium.

James E. Short is a lead scientist at the San Diego Supercomputer Center at the University of California, San Diego, in La Jolla, California.

Fredrik Svahn is an assistant professor of applied information technology at the University of Gothenburg in Gothenburg, Sweden, and is affiliated with the Swedish Center for Digital Innovation at Umeå University in Umeå, Sweden.

Steve Todd is fellow and vice president of strategy and innovation at Dell EMC, a part of Dell Technologies.

Leslie P. Willcocks is a professor of management at the London School of Economics and Political Science.

H. James Wilson is managing director of information technology and business research at the Accenture Institute for High Performance.

Barbara H. Wixom is a principal research scientist at the MIT Center for Information Systems Research.

Index

Accenture. *See* Nanterme, Pierre
Accountability, 47–48
Adobe Target, 78
Agility, 34, 186–187
AI (artificial intelligence), 75–80, 189, 190
Airbnb, 15, 17, 184
Allied Talent, 22–23
Amazon, 17, 29, 34, 109
Analytics. *See* HR analytics
Anaplan, 79
Apple, 8
Artificial intelligence (AI), 75–80, 189, 190
Associated Press (AP), 87, 94–95
Automating services, 81–100
  anxiety about, 81
  applications of, 82–83
  building capabilities, 97–100
  business operations as leading, 94
  cognitive automation, 83, 85
  command center for, 97–98
  customer/employee perceptions of, 94–97
  developing strategy for, 86–92
  executing, 93–97
  human-robot teamwork, 86, 89, 100
  IT's role in, 93–96
  multiple benefits of, 90–91
  robotic process automation, 82–86, 88–90, 101n7
  senior management's support of, 87–88, 90
  service characteristics, 83–86
  skill sets for, 98–100
  sourcing options for, 91–92, 101–101n11, 101n9
  sponsors, champions, managers of, 93–94
  for tasks within jobs vs. whole jobs, 96, 102n14–15
Aviso, 79

Baby boomers, 167
Barnes & Noble, 34
Beane, Billy, 68–69
Beiersdorf, 129

Biases in decision making, 67–69
Biotechnology-based therapies, 106–108
Bizible, 78
Blockbuster, 32–33
Blue Prism Group, 85
Boston Consulting Group, 3
BPO (business-process outsourcing), 92
Branding. *See under* social media, employee engagement on
Buford, R. C., 67
Burberry Group, 123
Business-process outsourcing (BPO), 92

Caesars Entertainment Operating, 50
Capital One, 48
Capital One Financial, 42
Cars, self-driving, 189
Cognitive computing, 189
Core employees, 23, 26–28
CRM (customer relationship management), 10
Crowdsourcing, 141–154
  advantages and challenges of, 142, 151–152
  defining problems for the problem solvers, 147–151
  new capabilities for the organization, 146, 151–154
  websites for, 143–147
C Space, 131
Customer relationship management (CRM), 10

Data analysts (quants) vs. decision makers, 65–66
Data hubris, 67
Data monetization, 39–48
  accessibility/quality of data, 47
  accountability, 47–48
  operational efficiency, 40–41
  overview of, 39–40
  selling data, 43–47, 48, 57
  wrapping information around products, 41–43, 47
Data quality, 148
Data translators, 66–71
Data valuation, 49–58
  activity value (data in use), 54
  asset/stock value, 52
  examples of data purchases, 49–50
  heuristic, 58
  in-house experts in, 57
  making policies explicit and sharable, 55–57
  needs-based, 51, 56–58
  research on, 50–52
  top-down vs. bottom-up, 57–58
  tracking, 56
Decision makers vs. data analysts (quants), 65–66
Defense Advanced Research Projects Agency (DARPA) Network Challenge, 156
Deloitte, 19
Digital innovation, 111–119
  embracing, 118–119
  external vs. internal collaboration, 115–116, 117
  flexibility vs. control, 116–118

new vs. established capabilities, 112–114, 117
process vs. product focus, 114–115, 117
Digital keys, 115
Digital natives, 167, 170
Digital transformation
chief transformation officer, xv
a company's need for, 14–15
definitions of, xi–xiii
emerging/disruptive technologies leveraged by, 15–16
executives' desire for, 17–18
getting started, xv–xvi
by industry disruptors, 17
myths about, 13–18
by profitable vs. failing companies, 16–17
resistance to, 16–17
Disruption, 17, 184, 190. *See also* technology, emerging/disruptive
Dollar Shave Club, 31
Domain experts, 66

Emotional bias, 68–69
Employees. *See* social media, employee engagement on; talent acquisition/retention
Enterprise resource planning (ERP), 10

Facebook, 128, 133, 136, 155, 160–161, 168
"Fail fast" mantra, xvi
FedEx, 42
Ferrara, Lou, 87
Focus, 187–188
Fortune 500 growth strategies, 29, 33

Gainsight, 79
General Motors, 29
Generation X, 167
Gene therapy, 107–108
Gillette, 31
Glassdoor, 170, 174
Global Comparative Performance Assessment Study, 125, 138n8
Google/Google+, 54, 160
Growth strategy, 29–35
acquisitions, 29
ambition, 33
analysis and invention, 31–32
catalyzing action, 33–35
defensive strategies, 29
dynamism and nimbleness, 30, 32–33
traditional, 30–31
Gustavsson, Mikael, 113

Harris, Del, 70
Harris, Jeanne G., 66
Hastings, Reed, 33
Hewlett-Packard, 79
HR analytics, 59–64
Human-technology interaction, 188–190, 192
Hyve Group, 131

IBM, 69
In-Car Delivery, 115
Incentives, 156–158
Information Systems Group, 85

Information technology (IT), 51, 106
InnoCentive, 26, 145
Interpretation gap, 66

Jet, 29
Johnson & Johnson, 48

Kaiser Permanente, 5–7, 10
Keyes, Jim, 32
Key performance indicators (KPIs), 34, 76
Kit Kat, 135

Lazer, David, 67
Leadership, xiii–xv. *See also* Nanterme, Pierre
LinkedIn, 49, 133–134
London School of Economics and Political Science, 85
L'Oréal, 160, 167, 170, 172
Loyalty, 5, 27, 50, 170
Lyft, 17, 29

Machine learning, 75–80, 189
Market Basket, 168
Massey, Cade, 68
Mehrotra, Vijay, 66
Mejdal, Sig, 69
Mentoring, reverse, 167
Microsoft, 47–48, 49
MIT Center for Information Systems Research, 3
MIT Red Balloon Challenge Team, 156
MIT Sloan Management Review, 19
Mobility, 189
Monetizing data. *See* data monetization
Monoclonal antibodies, 107
Moody's Investors Service, 49

Nadella, Satya, 40
Naím, Moisés, 155
Nanterme, Pierre
  on Accenture as a digitalization case study, 183–184
  on "digital first," 182
  on disruption, 184, 192
  on focus, 187–188
  on human-technology interaction, 188–190, 192
  on leaders, 182, 193
  on mobility, 189–190
  on navigation/agility, 186–187
  on productivity, 186
  on "the new," 183
navigation/agility, 186–187
Nestlé, 135
Netflix, 17, 32–33
Network measurability bias, 158
Nivea Invisible for Black & White, 129
Nutanix, 79

Objective and key results (OKRs), 34, 76
Olo, 22
Open innovation. *See* crowdsourcing
Operational backbone, 8–11
Operational efficiency, 40–41
Optimizely, 78
O2, 90–91, 94, 102n15
Overconfidence bias, 68

Pandora, 116–117
Patagonia, 159–160
Pernod Ricard, 159–160, 163
Pharmaceutical business model, 107–108
Pinterest, 126
Plant breeding case study. *See* crowdsourcing
Process leaders, 47
Product Development and Management Association (PDMA), 125, 138n8
Productivity, 186
Product leaders, 47–48

Residual analysis, 145
Reverse mentoring, 167
Robots. *See* artificial intelligence; automating services

Sales
- A/B tests, 77–78
- machine learning used in, 75–80
- reading social cues, 77

Salesforce, 78
Schick, 31
Schindler Group, 7–8
self-driving cars, 189
ShoreTel, 173–174
6sense, 77
Six Sigma, 34
Social media
- generating sustainable value on, 155–158
- generational differences among users, 167
- incentive structures on, 156–158
- mobilizations via, 155–156
- number of users, 159

Social media, employee engagement on, 159–175
- employee branding, fostering, 166–174
- employee branding, generally, 162–166
- employees as brand ambassadors, 159–160, 172
- employer vs. employee branding, 163
- and etiquette/confidentiality, 169
- implications for companies, 174–175
- "liking" their employers, 161, 169
- policing of, 174
- research on, 160–161, 164
- rewards for, 173–174
- training employees in, 169–170, 175

Social media's role in innovation, 123–137
- cocreating (collaborating), 127, 130–132
- communication skills, 127, 134–135
- customer insights, generating, 126–129
- dedicated strategy, 136–137
- exploring data for innovations, 128–130
- information source, 137–138n4
- innovation leaders, 137
- internal communication, 138n4

in mature vs. newer companies, 138n6
platform-specific strategies, 133–134
research on, 125–126, 138n8
socializing and relationships, 133
vs. traditional advertising, 134
types of social media, 128
types of users, 128
utilization issues, 123–125
Société Générale, 159–160
Sourcing. *See also* crowdsourcing
cloud, 92, 102n11
consulting, 91
insourcing, 91–92
outsourcing, 92
tool providers, 91, 101n9
Spotify, 116–117
State Street, 43, 45, 48
State Street Global Exchange, 43, 45
Strategic planning. *See* growth strategy
Strategy, developing, 3–11
choosing a single strategy, 5, 7–8
customer engagement strategies, 3, 5–6
digitized solutions strategies, 3, 5–7
operational backbone, 8–11
purpose of, 3–4
study of, 3
SurveyMonkey, 79
Syngenta, 141–145, 148, 153. *See also* crowdsourcing

Talent acquisition/retention, 19–28. *See also* crowdsourcing
core employees, 23, 26–28
cultivating on-demand talent markets, 24–26
employee network dynamics, 60–62
environment, 26–27
freelance talent market platforms, 20–22
full-time and part-time talent, balancing, 24, 26
hiring/firing policies, 60–62
on-demand talent, delegating to, 26–27
on-demand talent and core employees, blending, 23, 28
overview of, 19–20
via training and development, 22–23, 27–28
Technology, emerging/disruptive
vs. conventional, 15–16
disruption, 184, 190
human-technology interaction, 188–190, 192
invention vs. innovation/commercialization, 105–106
organizing for, 105–110
Telefónica, 90, 94, 102n15
3M, 27
Thulin, Inge, 27
Topcoder, 22
Twitter, 126, 128, 155, 169–170

Uber, 17, 22, 184
Uber Technologies, 15
Unilever, 31
USAA, 8

VHA, 95
Villanova University, 13–14
Volvo Cars, 111–119
Volvo Cloud, 115–116
Volvo On Call, 115

Walmart, 29
Waze, 54
Work Market, 20, 22, 24
Workplace, 168, 171
Wrapping, 41–43, 47

Xchanging, 82–84, 93–94, 97
Xerox, 106

Yammer, 168, 171
YouTube, 161